The Space Age

1957 The Soviet Union launches Sputnik 1, the first artificial satellite of Earth; Geoffrey Burbidge, E. Margaret Burbidge, William Fowler, and Fred Hoyle explain how the elements are made inside stars.

1958 Using the U.S.'s first satellite, Explorer 1, James Van Allen discovers the Earth's radiation belts ("magnetosphere").

1960 Frank Drake begins the Search for Extraterrestrial Intelligence at the National Radio Astronomy Observatory in Green Bank, West Virginia.

1961 Yuri Gagarin makes the first manned space flight.

1963 Valentina Tereshkova is the first woman in space.

1967 Jocelyn Bell and Anthony Hewish discover pulsars.

1969 Neil Armstrong and Buzz Aldrin walk on the Moon.

1979 In pictures sent from Voyager 1, Linda Morabito discovers erupting volcanoes on Jupiter's moon, Io.

1987 Ian Shelton discovers the first supernova, since 1604, plainly visible to the naked eye.

1990 Hubble Space Telescope launches.

1991 Alexander Wolszczan discovers planets orbiting a pulsar, the first known planets outside the solar system.

1995 Michel Mayor and Didier Queloz discover 51 Pegasi B, the first planet of a normal star beyond the Sun.

1998 Two teams of astronomers discover that the expansion of the universe seems to be getting faster, perhaps due to a mysterious force associated with the vacuum of space.

1999 Mars Global Surveyor gathers evidence suggesting that there was once a large ocean on Mars.

Famous Women in Astronomy

Caroline Herschel (1750–1848) Discovered eight comets.

Annie Jump Cannon (1863–1941) Devised basic method for classifying the stars.

Henrietta Swan Leavitt (1868–1921) Discovered method for measuring great distances in space.

Contemporary:

E. Margaret Burbidge Pioneered modern studies of galaxies and quasars.

Wendy Freedman Leader in measuring the expansion rate and age of the universe.

Sally Ride Trained as an astrophysicist, she was the U.S.'s first woman in space.

Nancy G. Roman NASA's first chief astronomer, she championed the development of telescopes in space.

Vera C. Rubin Investigated rotation of galaxies, detecting existence of dark matter.

Carolyn Shoemaker Discovered many comets including one that smashed into Jupiter.

Jill Tarter Leader of the largest search for extraterrestrial intelligence, Project Phoenix.

Astronomy For Dummies®

Cheat Sheet

An Astronomical Timeline

2000 B.C. According to legend, two Chinese astronomers are executed for not predicting an eclipse and for being drunk as it was underway.

129 B.C. Hipparchos completes the first catalog of the stars.

A.D. 150 Ptolemy publishes theory of Earth-centered universe.

970 al-Sufi prepares catalog of over 1,000 stars.

1420 Ulugh-Beg, prince of Turkestan, builds a great observatory and prepares tables of planet and star data.

1543 While on his deathbed, Copernicus publishes the theory that planets orbit around the Sun.

1609 Galileo discovers craters on Earth's Moon, moons of Jupiter, the turning of the Sun, and the presence of innumerable stars in the Milky Way by use of the telescope.

1666 Isaac Newton begins work on theory of universal gravitation.

1671 Newton demonstrates his invention of the reflecting telescope.

1705 Edmond Halley predicts that a great comet will return in 1758.

1758 On Christmas, farmer/amateur astronomer Johann Palitzch discovers the return of Halley's Comet.

1781 William Herschel discovers Uranus.

1791 Benjamin Banneker, first African-American scientist, begins star observations needed for the geographical survey to establish the future capital city of the United States, Washington, DC.

1833 Abraham Lincoln and thousands of others see enormous meteor shower over North America on November 12 through 13.

1842 Christian Doppler discovers the principle by which sound or light is shifted in frequency and wavelength due to the motion of its source with respect to the observer.

1846 Johann Galle discovers Neptune.

1916 Albert Einstein proposes the General Theory of Relativity, which explains the nature of gravity and the bending of light as it passes the Sun, and which in effect predicts the existence of black holes and the twisting of time and space in the vicinity of a massive, spinning object.

1923 Edwin Hubble proves that galaxies lie beyond the Milky Way.

1926 First launch of liquid-fuel rocket, developed by Robert Goddard.

1930 Clyde Tombaugh discovers Pluto.

1931 Engineer Karl Jansky discovers radio waves from space.

1939 Hans Bethe explains the energy source of the Sun and other stars.

1940 Engineer Grote Reber reports the first radio telescope survey of the sky.

Copyright © 1999 Wiley Publishing, Inc. All rights reserved.

Item 5155-8.

For more information about Wiley Publishing, call 1-800-762-2974.

For Dummies: Bestselling Book Series for Beginners

Praise for Astronomy For Dummies

"Steve Maran has long conveyed the excitement of astronomical discovery to ordinary folks. If you're just starting your journey through the cosmos — or even if you've already walked down that road — you need his travel guide by your side."

> — Leif J. Robinson, Editor-in-Chief, *Sky & Telescope*

"If you have always felt that astronomy is over your head, then *Astronomy For Dummies* is for you. In it, Steve Maran reminds you that while looking up can be fun, knowing a thing or two about what you see can be fun, too."

> — Neil deGrasse Tyson, Astrophysicist and Director, Hayden Planetarium, New York City

"You may think the cosmos is a vast and mysterious place, but *Astronomy For Dummies* will make it seem as friendly and familiar as your own backyard. Author Steve Maran is the perfect guide. He knows his way around the universe, and knows how to explain comets, planets, black holes, and the Big Bang — plus everything in between — in terms anyone can understand. If you always wondered what's really out there and what it all means, this is the book for you."

> — Michael D. Lemonick, author of *The Light at the Edge of the Universe* and *Other Worlds*

"One of the best known astronomers in the world, Steve Maran has written a marvelous, easy-to-follow introduction to the world above us. It's fun to read, easily understandable, and leaves you with the feeling that you've really become acquainted with the night sky. I wish I had this book when I was starting in astronomy."

> — David H. Levy, Science Editor, *Parade*

"With wit and good humor, Steve Maran here initiates the curious or confused into the mysteries of astronomy. A professional astronomer who often serves as press officer for the American Astronomical Society, Maran knows how to sniff out the significance in fast-breaking news stories, and he's superb in presenting complex ideas both simply and accurately."

> — Owen Gingerich, Harvard University

"Steve Maran is the quintessential astronomer . . . he is dazzled by the wonders of creation and committed to spreading the word to anyone willing to join him on a glorious tour of the cosmos. Maran is a researcher specializing in ultraviolet astronomy, yet his vantage point is not the highly statistical studies that dominate contemporary astronomy. It is that of a lone person standing on a hill on a moonless night, awestruck by the brilliant band of light that is our Milky Way galaxy.

Maran offers an exceptionally clear introduction to the 'language of light,' the ultimate guide to unlocking the secrets of the cosmos. He shows us how tiny beads of light, called photons, illuminate the violent realms of black holes and exploding stars, allows us to peer into the secret lives of comets and meteors, or visit the diverse landscapes of other planets. Maran takes us out into the incomprehensible scales of deep space, yet he urges readers to 'start with your eyes.' In *Astronomy For Dummies,* he offers a wonderful 'how-to' guide for that very human urge to crane your neck on a starry night and marvel at the amazing drama of creation."

— Thomas Lucas, Producer/Director, "Voyage to the Milky Way"
PBS 1999

™

...FOR DUMMIES

References for the Rest of Us!®

BESTSELLING BOOK SERIES

Do you find that traditional reference books are overloaded with technical details and advice you'll never use? Do you postpone important life decisions because you just don't want to deal with them? Then our *For Dummies*® business and general reference book series is for you.

For Dummies business and general reference books are written for those frustrated and hard-working souls who know they aren't dumb, but find that the myriad of personal and business issues and the accompanying horror stories make them feel helpless. *For Dummies* books use a lighthearted approach, a down-to-earth style, and even cartoons and humorous icons to dispel fears and build confidence. Lighthearted but not lightweight, these books are perfect survival guides to solve your everyday personal and business problems.

> *"More than a publishing phenomenon, 'Dummies' is a sign of the times."*
>
> — *The New York Times*

> *"...you won't go wrong buying them."*
>
> — *Walter Mossberg, Wall Street Journal, on For Dummies books*

> *"A world of detailed and authoritative information is packed into them..."*
>
> — *U.S. News and World Report*

Already, millions of satisfied readers agree. They have made For Dummies the #1 introductory level computer book series and a best-selling business book series. They have written asking for more. So, if you're looking for the best and easiest way to learn about business and other general reference topics, look to For Dummies to give you a helping hand.

Wiley Publishing, Inc.

Astronomy
FOR
DUMMIES®

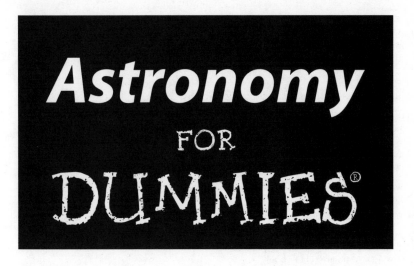

by **Stephen P. Maran, Ph. D.**

Wiley Publishing, Inc.

Astronomy For Dummies®

Published by
Wiley Publishing, Inc.
909 Third Avenue
New York, NY 10022
www.wiley.com

For general information on our other products and services or to obtain technical support, please contact our Customer Care Department within the U.S. at 800-762-2974, outside the U.S. at 317-572-3993, or fax 317-572-4002.

Wiley also publishes its books in a variety of electronic formats. Some content that appears in print may not be available in electronic books.

Library of Congress Cataloging-in-Publication Data:

Library of Congress Catalog Card No.: 99-66326

ISBN: 0-7645-5155-8

Manufactured in the United States of America

10 9 8 7

1O/TR/RQ/QS/IN

About the Author

Stephen P. Maran, Ph.D., a 30-year veteran of the space program, is the 1999 recipient of the Klumpke-Roberts Award of the Astronomical Society of the Pacific for "outstanding contributions to the public understanding and appreciation of astronomy." He received the NASA Medal for Exceptional Achievement in 1991 and was the A. Dixon Johnson Lecturer in Scientific Communication at Pennsylvania State University in 1990. He's also taught astronomy at the University of California, Los Angeles, and the University of Maryland, College Park. As press officer of the American Astronomical Society, he presides over media briefings that bring the news of astronomical discoveries to people worldwide.

Maran began practicing astronomy from rooftops in Brooklyn and a deserted golf course in the far reaches of the Bronx, and graduated to conduct professional research with telescopes at the Kitt Peak National Observatory in Arizona, National Radio Astronomy Observatory in West Virginia, Palomar Observatory in California, and Cerro Tololo Inter-American Observatory in Chile, as well as with instruments in space, including the Hubble Space Telescope and the International Ultraviolet Explorer. He's observed total eclipses of the Sun from the Gaspé Peninsula and elsewhere in Quebec, from Baja California in Mexico, and at sea off New Caledonia and Singapore, and in the eastern Pacific, as well as on shore in the good old U.S.A.

In the course of spreading the good word about astronomy, Maran has lectured on black holes in a bar in Tahiti, and explained an eclipse of the Sun on NBC's *Today* show. He's also spoken on eclipse and comet cruises of Cunard's *Queen Elizabeth 2* and *Vistafjord* and Sitmar Line's *Fairwind*. He's addressed audiences ranging from Seattle school children and Atherton, California, Girl Scouts to the National Academy of Engineering in Washington, D.C., and subcommittees of the U.S. House of Representatives and the U.N. Committee on the Peaceful Uses of Outer Space.

Maran is editor of *The Astronomy and Astrophysics Encyclopedia* and has co-authored or edited eight other books on those subjects, including a college textbook, *New Horizons in Astronomy,* and two compendiums of space program discoveries, *A Meeting with the Universe* and *Gems of Hubble*. He has written many articles for *Smithsonian* and *Natural History* magazines, and has served as a writer and consultant for the National Geographic Society and Time-Life Books.

He's a graduate of Stuyvesant High School in New York City, where he served a full season on the Math Team without incurring serious injury, and of Brooklyn College. His M.A. and Ph.D. in astronomy are both from the University of Michigan. Maran is married to Sally Ann Scott, a journalist. They have three children.

Dedication

To Sally, Michael, Enid, and Elissa with all my love.

Author's Acknowledgments

Thanks first to my family and friends who put up with me in the writing of this book. Thanks also to my agent, Skip Barker of the Wilson-Devereaux Company, who goaded and guided me in this project, and to Stacy Collins at Hungry Minds, Inc., for her faith in it.

I'm grateful to Ron Cowen and Dr. Seth Shostak for their contributions to this book, to Kathy Cox and Diane Smith who organized and edited it, and to their skilled colleagues on the editorial and production teams at Hungry Minds, who made the book better and brighter, and to Dr. Philip Plait, proprietor of the "Badastronomy" Web site, who ensured that bad information was winnowed out.

Thanks as well to artist Brent Pallas, and astronomical photographers Jerry Lodriguss and Dr. David Malin, who supplied some of their excellent photos, and to the organizations who kindly gave permission to use many of the other photographs in this book. I also thank the producer of the star maps, Robert Miller, and the producer of the planetary tables, Robert Victor.

About Our Guest Authors

Ron Cowen has been a writer at *Science News,* a weekly news magazine, since 1990. Since 1991, he has covered all aspects of astronomy for *Science News* as both reporter and editor. He has also contributed more than 30 stories to *The Washington Post,* ranging from backstage at the Nutcracker Ballet to a centennial celebration of the life of George Gershwin and the invention of the phonograph. He has also contributed articles to *Astronomy* magazine and *USA Today.*

Dr. Seth Shostak is the Public Programs Scientist at the SETI Institute, in Mountain View, California, where he participates in the search for cosmic company. He has degrees from Princeton University, and the California Institute of Technology. His career includes radio astronomy research on galaxies, teaching, writing, and the making of numerous popular science films. He also has an interest in railroads, and claims to be the inventor of the electrical banana.

Publisher's Acknowledgments

We're proud of this book; please send us your comments through our online registration form located at www.dummies.com/register.

Some of the people who helped bring this book to market include the following:

Acquisitions, Editorial, and Media Development

Project Editor: Kathleen M. Cox

Acquisitions Editor: Stacy S. Collins

Senior Copy Editor: Susan Diane Smith

Technical Editor: Philip Plait, Ph.D., www.badastronomy.com

Acquisitions Coordinator: Karen S. Young

Editorial Manager: Pamela Mourouzis

Editorial Assistant: Carol Strickland

Cover Photo: Hungry Minds, Inc.

Production

Project Coordinator: E. Shawn Aylsworth

Layout and Graphics: Amy M. Adrian, Kate Jenkins, Jill Piscitelli, Barry Offringa, Anna Rohrer, Doug Rollison, Brent Savage, Jacque Schneider, Maggie Ubertini, Dan Whetstine, Erin Zeltner

Proofreaders: Laura Albert, John Greenough, Betty Kish, Marianne Santy, Rebecca Senninger

Indexer: Rachel Rice

Special Help
Linda S. Stark, Jon Malysiak, Janet Withers, Donna Frederick

Publishing and Editorial for Consumer Dummies
Diane Graves Steele, Vice President and Publisher, Consumer Dummies
Joyce Pepple, Acquisitions Director, Consumer Dummies
Kristin A. Cocks, Product Development Director, Consumer Dummies
Michael Spring, Vice President and Publisher, Travel
Brice Gosnell, Publishing Director, Travel
Suzanne Jannetta, Editorial Director, Travel

Publishing for Technology Dummies
Richard Swadley, Vice President and Executive Group Publisher
Andy Cummings, Vice President and Publisher

Composition Services
Gerry Fahey, Vice President of Production Services
Debbie Stailey, Director of Composition Services

Contents at a Glance

Introduction 1

Part I: Stalking the Cosmos 7
Chapter 1: Seeing the Light: The Art and Science of Astronomy 9
Chapter 2: Skywatching: Join the Crowd 31
Chapter 3: The Way You Watch Tonight: Observing the Skies 43
Chapter 4: Checking Out Visitors: Meteors, Comets, and Man-Made Moons 59

Part II: Once Around the Solar System 77
Chapter 5: Earth and Its Moon 79
Chapter 6: Earth's Near Neighbors: Mercury, Venus, and Mars 99
Chapter 7: The Asteroid Belt and Near Earth Objects 117
Chapter 8: Jupiter and Saturn: Great Balls of Gas 125
Chapter 9: Far Out! Uranus, Neptune, and Pluto 135

Part III: Old Sol and Other Stars 143
Chapter 10: The Sun: Star of the Earth 145
Chapter 11: The Stars: Nuclear Reactors 169
Chapter 12: Galaxies: The Milky Way and Beyond 195
Chapter 13: Black Holes and Quasars 217

Part IV: The Remarkable Universe 229
Chapter 14: SETI and Planets of Other Suns 231
Chapter 15: Dark Matter and Antimatter 243
Chapter 16: The Big Bang and the Evolution of the Universe 251

Part V: The Part of Tens 261
Chapter 17: Ten Strange Facts about Astronomy and Space 263
Chapter 18: Ten Common Errors about Astronomy and Space 267

Part VI: Appendixes **271**

Appendix A: Finding the Planets: 2000 to 2004 273
Appendix B: Star Maps 285
Appendix C: Glossary 293

Index **297**

Cartoons at a Glance

By Rich Tennant

"I'm glad you were able to see a white dwarf and a red giant last night. I just hope you remembered not to stare."

page 143

page 7

"I think what you mean, dear, is a 'magnetic' storm."

page 77

"Along with 'Antimatter' and 'Dark Matter,' we've recently discovered the existence of 'Doesn't Matter,' which seems to have no effect on the universe whatsoever."

page 229

"Paul, turn off your flashlight. There's a real interesting star cluster I'm trying to get a picture of."

page 271

page 261

Fax: 978-546-7747
E-mail: richtennant@the5thwave.com
World Wide Web: www.the5thwave.com

Table of Contents

Introduction *1*

Who Do I Think You Are? 2
How This Book Is Organized: Your Journey through the Skies 2
    Part I: Stalking the Cosmos 3
    Part II: Once Around the Solar System 3
    Part III: Old Sol and Other Stars 3
    Part IV: The Remarkable Universe 4
    Part V: The Part of Tens 4
    Part VI: Appendixes 4
Icons Used in This Book 4
Where Do You Go from Here? 6

Part I: Stalking the Cosmos *7*

Chapter 1: Seeing the Light: The Art and Science of Astronomy 9

Astronomy: A Science of Observation 10
The Language of Light 11
    They wondered as they wandered: Planets versus stars 12
    If you see a Great Bear, start worrying: Naming stars and
        constellations 13
    Messier and other sky objects 21
    The smaller, the brighter: Getting to the root of magnitudes 21
    Don't make light of a light-year 23
    The fixed stars are moving all the time 24
Gravity: A Force to be Reckoned With 27
A Commotion of Motion 28

Chapter 2: Skywatching: Join the Crowd 31

All Eyes on the Skies: You Are Not Alone! 31
    Join an astronomy club for star-studded company 32
    Check out resources: Web sites, magazines, and more 33
Visit Observatories and Planetariums for Guided Viewing 35
    Observatories 36
    Planetariums 37
Vacation with the Stars: Star Parties, Eclipse Cruises, and Telescope
    Motels 38
    Star parties 38
    Eclipse cruises and tours: On the path of totality 39
    Telescope motels 41

Chapter 3: The Way You Watch Tonight: Observing the Skies 43
Begin with Naked-Eye Observation 44
Seeing Stars: A Primer on Sky Geography 46
As the Earth turns 47
Finding the North Star (Polaris) 47
For a Better View, Use Binoculars or a Telescope 49
Binoculars: Best for sweeping the sky 50
Telescopes: When closeness counts 53
A Plan for Dipping into Astronomy 58

Chapter 4: Checking Out Visitors: Meteors, Comets, and Man-Made Moons 59
Meteors: Wishing on a Shooting Star 59
Sporadic meteors, fireballs, and bolides 60
Meteor showers — sometimes the best sky sights of the year 62
Comets: The Lowdown on Dirty Ice Balls 67
Heads or tails: The structure of a comet 67
"Comets of the century" and all that 71
Your hunt for the Great Comet 72
Artificial Satellites: Astronomers Love 'em and Hate 'em 74
Observing artificial satellites 74
Finding satellite viewing predictions 75

Part II: Once Around the Solar System 77

Chapter 5: Earth and Its Moon 79
Earth: What's So Special About It? 80
Spheres of Influence: Dividing Earth 80
The magnetosphere: A main attraction 82
Spreading rock on the seafloor 83
Time and the Motions of Earth 83
Orbiting for all time 84
Tilting at the seasons 85
Estimating Earth's age: It's aeons and aeons! 87
Earth's Moon 88
Phases of the Moon 89
Eclipses of the Moon 90
Lunar geology 92
Giant impact: A theory of the origin of the Moon 95

Chapter 6: Earth's Near Neighbors: Mercury, Venus, and Mars 99
Hot, Shrunken, and Battered: Mercury's a Great Ball of Iron 99
Venus: Not a Nice Place to Live, or Even Visit 101

Mars: A Planet of Mysteries 102
Where has all the water gone? 102
Does Mars support life? 103
Observing the Terrestrial Planets 104
Understanding elongation, opposition, and conjunction 106
Identifying superior and inferior conjunctions 107
Viewing Venus and its phases 108
Watching Mars as it loops around 110
Outdoing Copernicus by observing Mercury 113
Discovering Why Earth Is Best: Comparative Planetology 115

Chapter 7: The Asteroid Belt and Near Earth Objects **117**
Asteroids: Leftovers from the Birth of the Solar System 117
Near Earth Objects: Is Earth Endangered? 119
When push comes to shove: Nudging an asteroid 120
Forewarned is forearmed: Surveying NEOs 121
Small Points of Light: Searching for Asteroids 122
Timing asteroidal occultations 123
Helping to track an occultation 123

Chapter 8: Jupiter and Saturn: Great Balls of Gas **125**
Inside Jupiter and Saturn: What You See Is Not What You Get 125
Observe Jupiter, by Jove! 126
In search of the Great Red Spot 127
Shooting for Galileo's moons 129
Ring Up Your Friends to See Saturn! 131
Cry "tilt" if you don't see the rings 133
Watch out for storms! 133
A moon of major proportions 133

Chapter 9: Far Out! Uranus, Neptune, and Pluto **135**
The Nature of Uranus and Neptune 135
The bull's-eye: Tilted Uranus and its rings and moons 136
Neptune and its backward moon 137
Pluto Is No Comic Character 137
Is Pluto a Planet? 139
What are Plutinos? 139
The Outer Planets: Viewing Challenges 140
Sighting Uranus 140
Distinguishing Neptune from a star 141
Striving to see Pluto 142

Part III: Old Sol and Other Stars *143*

Chapter 10: The Sun: Star of the Earth **145**
Don't Make Galileo's Blinding Mistake: Protect Your Sight
from the Sun 146

Surveying the Sunscape 146
        The Sun's size and shape: What keeps all those hot gases
            balled up together? 147
        The Sun's regions: Caught between the core and the corona 148
        The solar wind: Playing with magnets 150
        Solar activity and solar cycles: How's the weather out there? 151
        The solar neutrino mystery: Why are some missing? 154
        The Sun's lifespan: Will it ever die? 155
    Project or Filter to Protect: Safe Techniques for Solar Viewing 156
        Viewing the Sun by using projection 156
        Seeing the Sun through front-end filters 159
    Having Fun with the Sun: Solar Observation 161
        Tracking sunspots 161
        Looking at solar pictures on the World Wide Web 163
        Experiencing a total eclipse of the Sun 164

Chapter 11: The Stars: Nuclear Reactors **169**
    Life Cycles of the Hot and Massive 169
        Main sequence stars: A long adulthood 171
        Red giants 172
        Stars in the end states of stellar evolution 172
    Diagramming the Stars: Temperature, Mass, and Hertzsprung-
        Russell 178
        Spectral types: What color is my star? 178
        Star light, star bright: Classifying luminosity 180
        Mass determines class 180
        Interpreting the H-R diagram 181
    Born Together, Stay Together: Binary and Multiple Stars 182
        Where two or more gather 183
        The Doppler Effect and the importance of being binary 184
    Variable Stars 186
        Pulsating stars: Everybody's favorites 187
        Flare stars 188
        Exploding stars: supernovas and cataclysmic variables 189
        Eclipsing binary stars 190
        Microlensing events 191
    Stellar Neighbors to Know 192
    Helping Scientists by Observing the Stars 193

Chapter 12: Galaxies: The Milky Way and Beyond **195**
    Unwrapping the Milky Way: Earth's Galactic Home 195
        What shape is the Milky Way? 196
        Where can you find the Milky Way? 198
        How and when was the Milky Way formed? 199

Galactic Associates: Star Clusters 199
    Open clusters 199
    Globular clusters 201
    OB associations 202
Nebulae: Bright Shining Clouds and Some That Don't 203
    Planetary nebulae 204
    Supernova remnants 205
    Nebulae worth watching 206
Galaxies: Islands in the Universe 207
    Spiral, barred, and lenticular galaxies 208
    Elliptical galaxies 209
    Irregular, dwarf, and low surface brightness galaxies 210
    Great galaxies for gawkers 211
    The Local Group of Galaxies 212
    Clusters of galaxies 213
    Superclusters, great walls, and cosmic voids 214
Galactic Images on the World Wide Web 214

Chapter 13: Black Holes and Quasars **217**

Black Holes: Weird and Irresistible 217
    Types of black holes 218
    What's inside black holes? 218
    What's on the outside of black holes 221
    Distortions of space and time 222
Quasars: Defying Definitions 223
Active Galactic Nuclei: What in Blazes Are Blazars? 224

Part IV: The Remarkable Universe *229*

Chapter 14: SETI and Planets of Other Suns **231**

Is Anybody Out There? 231
SETI and Drake's Equation 232
Current SETI Searches: Listening for E.T. 234
    Project Phoenix 234
    Other SETI projects 237
    SETI searchers want you! 238
Hot Jupiters: The Truth about Extra-Solar Planets 238
    51 Pegasi's warm little world 239
    The Upsilon Andromedae system 240
    Planets suitable for life? 240
Continuing the Search 241

Chapter 15: Dark Matter and Antimatter 243

Dark Matter: The Glue That Holds Galaxies Together 243
The matter behind the missing mass 244
Debating the big question: What's the matter in dark matter? ...246
Searching for Dark Matter 247
WIMPS are shy but leave their mark 247
MACHOs make a brighter image 247
Dark matter can be mapped 248
Dark matter does matter 248
Antimatter: Opposites Attract 249

Chapter 16: The Big Bang and the Evolution of the Universe 251

Evidence for the Big Bang 252
Inflation: Swell Time in the Universe 253
Something from nothing: Inflation and the vacuum 254
Inflation and the shape of the universe 254
Funny Energy: Speeding Up Expansion? 255
Seeds of Galaxy Formation: A Closer Look at the Cosmic Microwave
Background 257
Hubble's Constant and the Age of the Universe 258
How fast do galaxies really move? 258
An inconstant constant? 259
How are galaxy distances measured? 259

Part V: The Part of Tens *261*

Chapter 17: Ten Strange Facts about Astronomy and Space 263

A Comet's Tail Often Leads the Way, Instead of Trailing Behind 263
Rocks from Mars Are All Over Earth 263
There Are Tiny Meteorites in Your Hair 264
You May Have Seen the Big Bang on an Old TV 264
Pluto Was Discovered from the Predictions of a False Theory 264
Sunspots Are Not Dark 265
On Venus, the Rain Never Falls on the Plain 265
The Ocean Tides Facing the Moon Are No Higher Than Tides on the
Other Side of Earth 265
A Star in Plain View May Have Erupted in an Enormous Supernova
Explosion, but No One Can Tell 265
Earth Is Made of Rare and Unusual Matter 266

Chapter 18: Ten Common Errors about Astronomy and Space 267

If You Were in the Asteroid Belt, You Would See Asteroids All
    Around You 267
Nuking a "Killer Asteroid" on a Collision Course for Our Planet
    Will Save Earth 267
Asteroids Are Round, Like Little Planets 268
The Big Bang Is Dead 268
A Meteorite That Just Fell to the Ground Is "Still Hot" 268
Summer Happens When the Earth Is Closest to the Sun 268
"The Light from That Star Took 1,000 Light-Years to Reach Earth" 269
When the Distance of a Galaxy Is Reported as, Say, "Two Billion
    Light-Years," That's a Fact 269
The "Morning Star" Is a Star 269
The Sun Is an Average Star 270
The Hubble Telescope Gets Up Close and Personal 270

Part VI: Appendixes *271*

Appendix A: Finding the Planets: 2000 to 2004 273

2000 274
2001 276
2002 278
2003 280
2004 282

Appendix B: Star Maps 285

Appendix C: Glossary 293

Index *297*

Introduction

Astronomy is the study of the sky, the science of cosmic objects and celestial happenings. It's the investigation of the nature of the universe we live in. Astronomers carry out the business of astronomy by looking and (for radio astronomers) listening. It's done with backyard telescopes, huge observatory instruments, and satellites orbiting Earth and looking out into space. Telescopes are sent up in sounding rockets and on unmanned balloons. And some instruments are sent into the solar system aboard deep space probes.

There's professional astronomy and amateur astronomy. About 13,000 professional astronomers engage in this science worldwide. Over 300,000 amateur astronomers live in the United States.

Professional astronomers do research on the Sun and the solar system, the Milky Way galaxy, and the universe beyond. They teach in universities, design satellites in government labs, and operate planetariums. They also write books, like this one. Most of them hold Ph.D.s, and nowadays — so many of them study abstruse physics or work with automated, robotic telescopes — they may not even know the constellations.

Amateur astronomers know the constellations. They share an exciting hobby. Some do it on their own, and thousands more join astronomy clubs and organizations of every description. The clubs pass on know-how from old hands to new members, share telescopes and equipment, and hold meetings where members tell what they've seen or photographed in the sky or hear lectures by visiting scientists.

Amateur astronomers also hold observing meetings where everyone brings their telescopes (or looks through someone else's 'scope). These observing sessions are conducted at regular intervals (such as the first Saturday night of each month) or on special occasions (such as the return of a major meteor shower each August, or the appearance of a bright comet like Hale-Bopp or Hyakutake). They save up for really big events, such as a total eclipse of the Sun, when thousands of amateurs and dozens of pros travel across Earth to position themselves in the path of totality to witness one of nature's greatest spectacles.

This book explains all you need to know to launch into the great hobby of astronomy. And it gives you a leg up on understanding the basic science of the universe as well. The latest space missions will make more sense: You'll understand why space probes are on the way to Saturn and its huge moon, Titan, right now, as well as en route to a comet to snag some dust from its tail.

You'll know why the Hubble Space Telescope peers out into space, and how to find out how other space missions are doing. And when the latest discoveries are reported in the newspapers and on TV, from space, from the big telescopes in Arizona, Hawaii, Chile, and California, or from other observatories around the world, you'll understand the background and appreciate the news.

Who Do I Think You Are?

You're probably reading this book because you want to know what's up in the sky, or what the scientists in the space program are doing. Perhaps you've heard that astronomy is a neat hobby, and you're wondering if it's for you and what equipment you would need.

You are not a scientist. You just enjoy looking at the night sky and have been captured by the knowledge bug, wanting to see and understand the real beauty of the universe.

You want to observe the stars, but you also want to know what it is you are seeing. Maybe you even want to make a discovery of your own. You don't have to be an astronomer to spot a new comet. You can even help listen for E.T. Whatever your goal, this book will help you achieve it.

Read only the parts you want, in any order you want. I try to explain what you need as you go.

Astronomy is fascinating and it's fun. So keep reading. Before you know it, you'll be pointing out Jupiter, spotting famous constellations and stars, or tracking the International Space Station as it whizzes by overhead. The neighbors will call you "stargazer". Police officers will ask you what you're doing in the park at night, or why you're up on the roof with those big binoculars. Tell 'em you're an astronomer. That's one they probably haven't heard.

How This Book Is Organized: Your Journey through the Skies

If you've already peeked at the table of contents, you know this book is divided into parts. Here's a brief description of what you can find in each of the six major parts.

Part I: Stalking the Cosmos

You see the stars night after night (well, not every night, but still. . .). You develop the same fascination that humans have always had with the cosmos. You watch, and you wonder, and you want to know more. What are those lights in the sky? What makes them look the way they do, and move the way they move. Are any of them dangerous? Should you be waving at your cosmic twin?

This part gets you started finding your own answers to some of these questions, based on the answers that have already been found. Thousands of amateur astronomers gather together to support each other and share their knowledge. Astronomy is fun, as well as practical (and, yes, even educational).

In this part, I give you pointers on observing sky objects with and without optics, selecting binoculars and telescopes, and positioning yourself for the best view. I introduce you to delightful cosmic visitors, and set you up to continue exploring the universe.

Part II: Once Around the Solar System

It's only natural to want to meet your neighbors. Earth's neighbors are a collection of planets, moons, and planetary debris linked in orbit around the Sun. Like all neighbors, they share certain characteristics, but they are all wildly different, too.

These chapters focus on the observational aspects of the planets, so that you'll know what you're seeing and can appreciate the view. Nevertheless, I also try and answer the Big Question: Is anything alive out there? And the answer so far is no. But we're still looking. And you may someday be the one who finds the definitive answer.

Part III: Old Sol and Other Stars

Wondering about galaxies far, far away? This part starts with the Sun and takes you through the stars, introducing you to red giants and white dwarfs, dropping in on distant galaxies and exotic sky objects, and ending with black holes. Do you really want to go there? You might get swept away.

But as the late, great Carl Sagan said, we are all star stuff. So understanding the stars and enjoying their diversity enhances our own connections to the stuff of the universe.

This part points out the best and the brightest of the sky objects for your observational pleasure. I also spell out the life cycles of the stars so you can appreciate the forces that power the universe and make it endlessly intriguing.

Part IV: The Remarkable Universe

Read this part when you need a diversion, something to stir your mind with thought-provoking ideas and possibilities. Curl up with a cup of cider and read about SETI, the search for extraterrestrial intelligence. Have scientists found any evidence that those little green men are out there? Find out about dark matter and antimatter (yes, antimatter exists in the real world, not just in science fiction). And, when you are ready, ponder the entire universe: How did it begin, what shape is it, and what's going to happen to it?

Part V: The Part of Tens

Did you ever find yourself at a social gathering, desperately trying to think of something unique and interesting to say? You searched your brain for some crowd-grabbing insight that would make everyone in the room take notice of your intelligence. Well, read this part, and you'll be ready for the next lapse in conversation. I offer you ten strange facts about space that are guaranteed to garner interest. And then I fill you in about ten major mistakes that people and the media have often made — and continue to make — when they discuss astronomy.

Part VI: Appendixes

The appendixes in this part offer information that will enhance your sky-watching experiences for years to come. The first appendix offers maps to help you find interesting stars. The second one gives you tables showing the approximate locations — at any time from the year 2000 through 2004 — of the four bright planets that are the most commonly observed: Venus, Mars, Jupiter, and Saturn.

Icons Used in This Book

Throughout this book, helpful little icons highlight particularly useful information — even if they're just telling you to not sweat the tough stuff. Here's what each symbol means.

This target puts you right on track to make use of some inside information as you pursue or get set to start your skywatching.

Observation is the key to astronomy, and these tips will help make you a pro at it. This helps you scope out techniques and opportunities to fine-tune your eyes.

Sometimes you've got to talk the talk in order to impress your friends and take your stargazing activities to new heights. This little guy lets you know when you have the right word.

How much trouble can you get into just watching the stars? Not much, if you're careful. But some things you can't be too careful about. This bomb tells you to pay attention so you don't get burned.

Don't be misled. This icon alerts you to the truth behind common reports and assumptions about astronomy.

This nerdy guy appears beside discussions that aren't critical if you just want to know the basics and start watching the skies. The scientific background can be good to know, but many people happily enjoy their stargazing without knowing a lot about the physics of supernovas, the mathematics of galaxy chasing, and just what is that funny energy anyway? This icon lets you know what you're in for.

Astronomy and space are rich in resources on the World Wide Web. You can stay up-to-date on just about any topic in astronomy and space science by selecting from among these sites. Web site addresses are notorious for rapid change. These URLs were accurate at press time, but don't be surprised if a few of them change over time.

Some techniques and equipment in astronomy and space science give the most bang for the buck. My favorites are flagged with this mark, which means that they're A-OK.

Where Do You Go from Here?

You can start off anywhere you want to. Worried about the fate of the universe? Start off with the Big Bang.

But it's more likely that you want to know what's in store for you as you pursue your passion for the stars.

Wherever you start, I hope you continue your cosmic exploration and the enjoyment, excitement, and enchantment people have always found in the skies.

Part I
Stalking the Cosmos

The 5th Wave
By Rich Tennant

In this part . . .

Humans have always been fascinated by objects and events in the sky. Throughout history, the interest in astronomy has been both practical and devotional. People navigated by the stars and planted crops according to the moon phases. They also built ritual sites (such as Stonehenge) and developed ceremonies to celebrate astronomical events. And folks wondered about the nature of objects in the heavens.

You can join this grand human tradition. In this part, I introduce the science of astronomy, and I offer techniques and advice for observing planets, comets, meteors, and other sights in the night sky.

Chapter 1

Seeing the Light: The Art and Science of Astronomy

In This Chapter

▶ Understanding the nature of astronomy

▶ Traveling light years and beyond

▶ Weighing in on gravity

Step outside on a clear (cloud-free) night and look at the sky. If you're a city dweller, or live in a close-in suburb, you see dozens, maybe hundreds, of twinkling stars. Depending on the time of the month, you may also see the Moon, and up to five of the nine planets that revolve around the Sun.

A "shooting star" flashes overhead — that's a meteor, a tiny bit of comet dust streaking through the upper atmosphere.

Much more slowly, a pinpoint of light moves steadily across the sky. Is it a space satellite, such as the Hubble Space Telescope, or just a high-altitude airliner? If you have a pair of binoculars, you may be able to see the difference.

The airliner may have running lights, and its shape may be perceptible.

If you're way out in the country — at the seashore away from resorts and developments, or in the mountains far from any floodlit ski slope — you see thousands of stars. The Milky Way is a beautiful pearly swath across the heavens. It's actually the cumulative glow from millions of faint stars, individually indistinguishable with the naked eye. At a really great observation place, such as Cerro Tololo in the Chilean Andes, you can see even more stars. And they hang like brilliant lamps in a coal black sky, often not even twinkling, just like in one of Van Gogh's *Starry Night* paintings.

When looking at the sky, you're actually practicing astronomy — observing the universe that surrounds you and trying to make sense of what you see.

Astronomy: A Science of Observation

Astronomy is the study of the sky, the science of cosmic objects and celestial happenings. It's the investigation of the nature of the universe we live in. Professional astronomers carry out the business of astronomy by looking and (for radio astronomers) "listening." They use backyard telescopes, huge observatory instruments, and satellites that orbit Earth and collect forms of light (such as ultraviolet radiation) that the atmosphere blocks from reaching the ground. Telescopes are sent up in sounding rockets (rockets equipped with instruments for making high-altitude scientific observations) and on unmanned balloons. And some instruments are sent out into the solar system aboard deep space probes.

Professional astronomers study the Sun and the solar system, the Milky Way galaxy, and the universe beyond. They teach in universities, design satellites in government labs, and operate planetariums. They also write books, like this one. Most of them hold Ph.D.s. So many of them study abstruse physics or work with automated, robotic telescopes that they have moved far beyond the observable night sky. They may not even know the *constellations* (groups of stars given names such as Ursa Major, the Great Bear, by ancient stargazers) that are many people's first introduction to astronomy. (You may already be familiar with the Big Dipper, an *asterism* in Ursa Major. An asterism is a named star pattern that is not one of the 88 recognized constellations. Figure 1-1 shows the Big Dipper in the night sky.)

Figure 1-1:
A photo
of the
Big Dipper.

In addition to the more than 13,000 professional astronomers worldwide, thousands of amateur astronomers enjoy studying the skies, more than 300,000 of them in the United States alone. Many amateurs also make useful scientific contributions.

Amateur astronomers usually know the constellations. They learn them first as guideposts to exploring the sky by eye, with binoculars, and with telescopes.

For thousands of years, everything people knew about the heavens was deduced by simply looking at the sky. So the first thing you really need to understand about astronomy is that almost everything it deals with

✔ Is learned by studying the light that comes to us from objects in space

✔ Is seen from a distance

✔ Is moving through space under the influence of gravity

This chapter introduces you to these concepts of light, distance, and gravity.

The Language of Light

Light brings us information about the planets, moons, and comets in our solar system; the stars, star clusters, and nebulae in our galaxy; and the objects beyond.

What astronomy isn't

Astronomy is not astrology! Nothing makes an astronomer madder than when someone innocently calls him or her "an astrologer." Astronomers believe that when Jupiter aligns with Mars, it's a beautiful spectacle for stargazers, not an omen of good or bad fortune.

Astronomers are not UFOlogists; they're not engaged in the search for unidentified flying objects (UFOs). *Usually,* they can identify what they're looking at. Both astronomers and UFOlogists look at the sky. Both see stars and planets. But generally speaking, only UFOlogists take seriously close encounters with supposed alien spacecraft or beings.

SETI, the Search for Extraterrestrial Intelligence, is another matter. Astronomers do conduct that program. They operate sensitive radio telescopes, listening for any hint of reproducible signals from the cosmos that appear to be deliberate transmissions from the planets of stars beyond our solar system. And, recently, they've begun looking for messages that may be arriving in flashes of light, possibly from powerful lasers operated by civilizations more advanced than our own.

Astronomers haven't heard from ET yet, but we're listening. Everything astronomers have learned about planets and stars leads most of us to believe that habitable planets exist elsewhere. Many astronomers believe, as the late Carl Sagan was fond of saying, "We are not alone."

In ancient times, folks didn't think about the physics and chemistry of the stars; they absorbed and passed down folk tales and myths: the Great Bear, the Demon star, the Man in the Moon, the dragon eating the Sun during a solar eclipse, and more. The tales varied from culture to culture. But many people did learn to recognize the patterns of the stars. In Polynesia, skilled navigators rowed across hundreds of miles of open ocean with no landmarks in view and no compass. They sailed by the stars, the Sun, and their knowledge of prevailing winds and currents.

Gazing at the light from a star, even the ancients noted its brightness, position on the sky, and color. This information helps people distinguish one sky object from another, and recognize them when they become old friends. Some basics of recognizing and describing what you see in the sky are

- Distinguishing stars versus planets
- Identifying constellations and stars by name
- Observing brightness (given as magnitudes)
- Charting sky position (measured in special units)
- Recognizing meteors and comets

They wondered as they wandered: Planets versus stars

The term planet comes from the ancient Greek *planetes* or wanderer. The Greeks, along with just about everyone else, noticed that five spots of light moved across the pattern of stars in the sky. Some moved steadily ahead; others occasionally looped back on their own paths. Nobody knew why. And these spots of light generally didn't twinkle, while the stars generally did. There wasn't a good explanation for that difference either. Every culture had a name for those five spots of light, or planets. Today, we call them Mercury, Venus, Mars, Jupiter, and Saturn. And almost everyone knows they aren't wandering through the stars; they are orbiting around the Sun, their central star.

Today we know that the planets are objects that can be smaller or bigger than Earth, but they're all much smaller than the Sun. They are all much nearer to us on the planet Earth than to stars other than the Sun, so planets have perceptible disks, at least through a telescope — that is, planets have a definite round shape and a discernable size. The stars are so far away from Earth that even as viewed through a powerful telescope, they're just points of light.

From mythology to science

After the Dark Ages, scientific explanations for things in the sky started to replace myths. Instead of the ancient Egyptian myth (for example) that the Sun and Moon were carried around the sky on the back of the goddess Nut, astronomers realized that the Earth turns, it orbits around the Sun, and the Moon orbits around Earth.

Isaac Newton formulated the theory of gravity, and people began to understand what keeps objects in orbit and why the planets at greater distances from the Sun complete their orbits at a slower pace than those closer in.

Spectrographs and other instruments were built and put on telescopes. These devices tell astronomers how hot the stars are, what substances are present in them, how fast they are moving toward or away from Earth, and other basic physical information. If they have magnetic fields, we can measure them from a distance. And we can estimate the strength of gravity on the surface of a star, the density of its gas, and more. (The word *gas* here refers to matter in a certain physical state, as in gas versus liquid, not to a particular gas. On a star, iron is a gas.)

About the hardest physical information to come by has been the distances of the stars and other objects beyond the planets of our solar system. Some stars look bright but are really garden-variety stars that just happen to be nearby (nearby means four light-years or more, but not hundreds of light-years; for a definition of *light-year* as a unit of distance, see the section "Don't make light of a light-year," later in this chapter). Other stars are so dim you need a powerful observatory telescope to see them, but they're right in our own back-yard. (Well, they're a mere dozen or two light-years away.)

If you see a Great Bear, start worrying: Naming stars and constellations

I used to tell planetarium audiences who were craning their necks to look at stars projected above them, "If you can't see a Great Bear up there, don't worry. Maybe those who *do* see a Great Bear should worry."

The ancients divided the sky into imaginary figures, such as Ursa Major (Latin for Great Bear); Cygnus, the Swan; Andromeda, the Chained Lady; and Perseus, the Hero. Each figure was identified with a pattern of stars. The truth is, to most people, Andromeda doesn't look much like a chained lady, or anything else (see Figure 1-2).

Today, the sky is divided into 88 constellations, which contain all the stars that you can see. The International Astronomical Union, which governs the science, set boundaries for the constellations so astronomers can agree on which star is in which constellation. Previously, sky maps drawn by different astronomers often disagreed. When you read that the Tarantula nebula is in Dorado (see Chapter 12), you know that to see this nebula, you must seek it in the Southern Hemisphere constellation Dorado, the Goldfish.

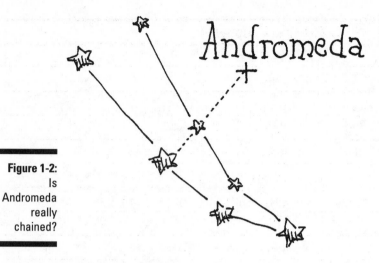

Figure 1-2:
Is
Andromeda
really
chained?

The largest constellation is Hydra, the Water Snake. The smallest is Crux, the Cross, which everybody calls the "Southern Cross." There's a Northern Cross, too, but you won't find it in a list of constellations; it's an asterism within Cygnus, the Swan. Although there's general agreement on the names of the constellations, there's no consensus on what each name means. For example, some astronomers call Dorado, the "Swordfish," but I'd like to spike that name. One constellation, Serpens, the Serpent, is broken into two sections, which are not connected. The two sections, which are located on either side of Ophiuchus, the Serpent Bearer, are Serpens Caput (the Serpent's Head) and Serpens Cauda (the Serpent's Tail).

The individual stars in a constellation often have no relation to each other except for proximity on the sky as visible from Earth. In space, the stars in a constellation may be completely unrelated to one another, with some being relatively near to Earth and some located at much greater distances in space. But they make a simple pattern for observers on Earth to enjoy.

As a rule, each of the brighter stars in a constellation was assigned a Greek letter, either by the ancient Greeks or by astronomers of later civilizations. In each constellation, the brightest star was supposed to be labeled alpha, the first letter of the Greek alphabet. The next brightest star was beta, the second Greek letter, and so on down to omega, the twenty-fourth and last Greek letter. (These are all lowercase Greek letters, by the way, not capitals, so they are written α, ß, . . . ω.)

So Sirius, the brightest star in the night sky — which is in Canis Major, the Great Dog — is called Alpha Canis Majoris. (Astronomers add a suffix here or there to put star names in the Latin genitive case — scientists have always liked Latin.) Table 1-1 is a list of the Greek alphabet, in order, with the names of the letters and their corresponding symbols.

Table 1-1	The Greek Alphabet
Letter	*Name*
α	Alpha
ß	Beta
γ	Gamma
δ	Delta
ε	Epsilon
ζ	Zeta
η	Eta
θ	Theta
ι	Iota
κ	Kappa
λ	Lambda
μ	Mu
ν	Nu
ξ	Xi
ο	Omicron
π	Pi
ρ	Rho
σ	Sigma
τ	Tau
υ	Upsilon
φ	Phi
χ	Chi
ψ	Psi
ω	Omega

When you actually look at the constellations today, you'll find that there are many exceptions to the rule that the order of brightness of the stars is the same as the Greek letters they are named on a star map. The exceptions exist because

✔ The letters were assigned based on naked-eye observations of brightness, which were not very accurate.

✔ Over the years, constellation boundaries were changed by star-atlas authors, so some stars were moved into a new constellation after that constellation's stars were already lettered.

✔ Many small and Southern Hemisphere constellations were drawn up long after the Greek period, and the practice wasn't always followed.

✔ Some stars have changed in brightness over the many centuries since Greek times.

A good (or bad) example is the constellation Vulpecula, the Fox, where only one of the stars (alpha) has a Greek letter.

When you look at a star atlas, you'll find that the individual stars in a constellation are not marked α Canis Majoris, ß Canis Majoris, and so on. Usually, the area of the whole constellation is marked 'Canis Major' and the individual stars are labeled just α, ß, and so on. When you read about a star in a list of objects to observe, say in an astronomy magazine, it probably won't be written in the style of Alpha Canis Majoris or even α Canis Majoris. Instead, to save space, it will be printed α CMa; "CMa" is the three-letter abbreviation for Canis Majoris. (It's also the abbreviation for Canis Major.) The abbreviation for each of the constellations is given in Table 1-2.

Astronomers didn't have special names such as Sirius for every star in Canis Major, so they just named them with Greek letters or other symbols. In fact, some constellations don't have a single named star. (Don't fall for those advertisements that offer to name a star for a fee. The International Astronomical Union doesn't recognize purchased star names.) In other constellations, Greek letters were assigned, but there were more than 24 readily visible stars, so there weren't enough Greek letters for them. So astronomers gave many stars numbers and letters from the Roman alphabet. Some examples of such star names are 236 Cygni, b Vulpeculae, HR 1516, and worse. There are even stars named RU Lupi and SX Sex. (I'm not making this up.) But like any other star, you can recognize them by their positions on the sky (as tabulated in star lists) and their brightness, colors, or other properties, if not their names.

Because alpha isn't always the brightest star in a constellation, another term is needed to describe that exalted status. *Lucida* is the word. The lucida of Canis Major is Sirius, the alpha star, but the lucida of Orion, the Hunter, is Rigel, which is Beta Orionis, and the lucida of Leo Minor, the Little Lion (a particularly inconspicuous constellation) is just 46 Leo Minoris.

Table 1-2 lists the 88 constellations, the brightest star in each, and the magnitude of that star. *Magnitude* is a measure of a star's brightness. (I talk about magnitudes a little later in this chapter in the section "The smaller, the brighter: Getting to the root of magnitudes.") When the lucida of a constellation is the alpha star and it has a name, I just list that name. For example, in Auriga, the Charioteer, the brightest star, Alpha Aurigae, is Capella. But when the lucida is not alpha, I give its Greek letter or other designation in parenthesis. For example, the lucida of Cancer, the Crab, is Al Tarf, which is Beta Cancri.

Table 1-2		The Constellations and Their Brightest Stars		
Name	*Abbreviation*	*Meaning*	*Star*	*Magnitude*
Andromeda	And	Chained Lady	Alpheratz	2.1
Antlia	Ant	Air Pump	Alpha Antliae	4.3
Apus	Aps	Bird of Paradise	Alpha Apodis	3.8
Aquarius	Aqr	Water Bearer	Sadalmelik	3.0
Aquila	Aql	Eagle	Altair	0.8
Ara	Ara	Altar	Beta Arae	2.9
Aries	Ari	Ram	Hamal	2.0
Auriga	Aur	Charioteer	Capella	0.1
Bootes	Boo	Herdsman	Arcturus	−0.04
Caelum	Cae	Chisel	Alpha Caeli	4.5
Camelopardalis	Cam	Giraffe	Beta Camelopardalis	4.0
Cancer	Cnc	Crab	Al Tarf (Beta Cancri)	3.5
Canes Venatici	CVn	Hunting Dogs	Cor Caroli	2.8
Canis Major	CMa	Great Dog	Sirius	−1.5
Canis Minor	CMi	Little Dog	Procyon	0.4
Capricornus	Cap	Goat	Deneb Algedi (Delta Capricorni)	2.9
Carina	Car	Ship's Keel	Canopus	−0.7
Cassiopeia	Cas	Queen	Schedar	2.2
Centaurus	Cen	Centaur	Rigil Kentaurus	−0.3
Cepheus	Cep	King	Alderamin	2.4

(continued)

Table 1-2 *(continued)*

Name	Abbreviation	Meaning	Star	Magnitude
Cetus	Cet	Whale	Deneb Kaitos (Beta Ceti)	2.0
Chamaeleon	Cha	Chamaeleon	Alpha Chamaeleontis	4.1
Circinus	Cir	Compasses	Alpha Circini	3.2
Columba	Col	Dove	Phakt	2.6
Coma Berenices	Com	Berenice's Hair	Beta Comae Berenices	4.3
Corona Australis	CrA	Southern Crown	Alpha Coronae Australis	4.1
Corona Borealis	CrB	Northern Crown	Alphekka	2.2
Corvus	Crv	Crow	Gienah (Gamma Corvi)	2.6
Crater	Crt	Cup	Delta Crateris	3.6
Crux	Cru	Southern Cross	Acrux	0.7
Cygnus	Cyg	Swan	Deneb	1.3
Delphinus	Del	Dolphin	Rotanev (Beta Delphini)	3.6
Dorado	Dor	Goldfish	Alpha Doradus	3.3
Draco	Dra	Dragon	Thuban	3.7
Equuleus	Equ	Little Horse	Kitalpha	3.9
Eridanus	Eri	River	Achernar	0.5
Fornax	For	Furnace	Alpha Fornacis	3.9
Gemini	Gem	Twins	Pollux (Beta Geminorum)	1.1
Grus	Gru	Crane	Alnair	1.7
Hercules	Her	Hercules	Ras Algethi	2.6
Horologium	Hor	Clock	Alpha Horologii	3.9
Hydra	Hya	Water Snake	Alphard	2.0
Hydrus	Hyi	Little Water Snake	Beta Hydri	2.8

Name	Abbreviation	Meaning	Star	Magnitude
Indus	Ind	Indian	Alpha Indi	3.1
Lacerta	Lac	Lizard	Alpha Lacertae	3.8
Leo	Leo	Lion	Regulus	1.4
Leo Minor	LMi	Little Lion	Praecipua (46 Leo Minoris)	3.8
Lepus	Lep	Hare	Arneb	2.6
Libra	Lib	Scales	Zubeneschemali (Beta Librae)	2.6
Lupus	Lup	Wolf	Alpha Lupus	2.3
Lynx	Lyn	Lynx	Alpha Lyncis	3.1
Lyra	Lyr	Lyre	Vega	0.0
Mensa	Men	Table	Alpha Mensae	5.1
Microscopium	Mic	Microscope	Gamma Microscopii	4.7
Monoceros	Mon	Unicorn	Beta Monocerotis	3.7
Musca	Mus	Fly	Alpha Muscae	2.7
Norma	Nor	Level and Square	Gamma Normae	4.0
Octans	Oct	Octant	Nu Octantis	3.8
Ophiuchus	Oph	Serpent Bearer	Rasalhague	2.1
Orion	Ori	Hunter	Rigel (Beta Orionis)	0.1
Pavo	Pav	Peacock	Alpha Pavonis	1.9
Pegasus	Peg	Winged Horse	Enif (Epsilon Pegasi)	2.4
Perseus	Per	Hero	Mirphak	1.8
Phoenix	Phe	Phoenix	Ankaa	2.4
Pictor	Pic	Easel	Alpha Pictoris	3.2
Pisces	Psc	Fish	Eta Piscium	3.6
Pisces Austrinus	PsA	Southern Fish	Fomalhaut	1.2
Puppis	Pup	Ship's Stern	Zeta Puppis	2.3
Pyxis	Pyx	Compass	Alpha Pyxidus	3.7

(continued)

Table 1-2 *(continued)*

Name	*Abbreviation*	*Meaning*	*Star*	*Magnitude*
Reticulum	Ret	Net	Alpha Reticuli	3.4
Sagitta	Sge	Arrow	Gamma Sagittae	3.5
Sagittarius	Sgr	Archer	Kaus Australis (Epsilon Sagittarii)	1.9
Scorpius	Sco	Scorpion	Antares	1.0
Sculptor	Scl	Sculptor	Alpha Sculptoris	4.3
Scutum	Sct	Shield	Alpha Scuti	3.9
Serpens	Ser	Serpent	Unukalhai	2.7
Sextans	Sex	Sextant	Alpha Sextantis	4.5
Taurus	Tau	Bull	Aldebaran	0.9
Telescopium	Tel	Telescope	Alpha Telescopium	3.5
Triangulum	Tri	Triangle	Beta Trianguli	3.0
Triangulum Australe	TrA	Southern Triangle	Alpha Trianguli Australis	1.9
Tucana	Tuc	Toucan	Alpha Tucanae	2.9
Ursa Major	UMa	Great Bear	Alioth (Epsilon Ursae Majoris)	1.8
Ursa Minor	UMi	Little Bear	Polaris	2.0
Vela	Vel	Sails	Suhail al Muhlif (Gamma Velorum)	1.7
Virgo	Vir	Virgin	Spica	1.0
Volans	Vol	Flying Fish	Gamma Volantis	3.6
Vulpecula	Vul	Fox	Anser	4.4

Identifying stars would be a lot easier if they had little name tags, like attendees at a convention, that you could see through your telescope but at least they don't have unlisted numbers.

Messier and other sky objects

Naming stars was an easy task. But what about all those other objects in the sky — galaxies, nebulae, star clusters, and the like (which I cover in Part III). Charles Messier, a French astronomer in the late 18th century, created a list of about 100 fuzzy sky objects, giving them numbers. His list came to be known as the *Messier Catalog,* and when you see the Andromeda Galaxy called by its scientific name, M31, you'll know what it means. Today, there are 110 objects in the standard *Messier Catalog.*

You can find pictures of the Messier objects and a complete list of them at The Messier Catalog Web site of Students for the Exploration and Development of Space at `www.nerdnet.nl/~angelo/phoenix/messier/Messier.html`. And you can find out how to earn a certificate for viewing Messier objects from the Astronomical League Messier Club Web site at `www.astroleague.org/al/obsclubs/messier/mess.html`.

Experienced amateur astronomers often engage in Messier marathons in which each person tries to observe every object in the *Messier Catalog* during a single long night. But in a marathon, you have no time to enjoy an individual nebula, star cluster, or galaxy. I say: "Take it slow" and savor their individual visual delights. A wonderful book on the Messier objects, which includes hints on how to observe each one, is Stephen J. O'Meara's *The Messier Objects* (Cambridge University Press and Sky Publishing Corporation, 1998).

There are thousands of other *deep sky objects*, the term amateurs use for star clusters, nebulae, and galaxies to distinguish them from stars and planets. You'll find many of them listed in viewing guides and sky maps by their NGC *(New General Catalogue)* and IC *(Index Catalogue)* numbers. For example, the bright double cluster, in Perseus, the Hero, consists of NGC 869 and NGC 884.

The smaller, the brighter: Getting to the root of magnitudes

A star map, constellation drawing, or list of stars always indicates each star's magnitude. The *magnitudes* are just a brightness scale. One of the ancient Greeks, Hipparchos (also spelled Hipparchus, but he wrote it in Greek) divided up all the stars he could see into six classes. The brightest stars were called magnitude 1 or *1st magnitude,* the next brightest bunch were the *2nd magnitude* stars, and so on, down to the dimmest ones, which were *6th magnitude.*

Notice that contrary to most common measurement scales and units, the brighter the star, the smaller the magnitude. The Greeks weren't perfect, however; even Hipparchos had an Achilles heel: He didn't leave room in his system for the very brightest stars, when accurately measured.

So, today, we recognize a few stars as having 0 magnitude, or even negative magnitude. Sirius, for example, is magnitude –1.5. And the brightest planet, Venus, is sometimes magnitude –4 (the exact value differs, depending on the distance Venus is from Earth at the time, and the direction it is with respect to the Sun).

Another omission: The Greeks didn't have a magnitude class for stars that they couldn't see. This didn't seem like an oversight at the time, because nobody knew about these stars. But today we know that millions of stars exist beyond our naked-eye view, and they all have magnitudes, too. Their magnitudes are larger numbers: 7 and 8 for stars easily seen in binoculars, and 10 or 11 for stars easily seen in a small but good telescope. The magnitudes reach as high (dim) as 21 for the faintest stars in the Palomar Observatory Sky Survey, and 30 or maybe 31 for the faintest objects imaged with the Hubble Space Telescope.

TECHNICAL STUFF

The mathematics of brightness

The 1st magnitude stars are about 100 times brighter than the 6th magnitude stars. In particular, the 1st magnitude stars are about 2.512 times brighter than the 2nd magnitude stars, which are about 2.512 times brighter than the 3rd magnitude stars, and so on. You mathematicians out there will recognize this as a systematic progression. Each magnitude is the 5th root of 100 (meaning that when you multiply a number by itself four times — for example, 2.512 x 2.512 x 2.512 x 2.512 x 2.512 — the result is 100). If you doubt my word and do this calculation, you will get a slightly different answer because I left off some decimal places.

Thus, you can calculate how faint a star is — compared to some other star — from its magnitude. If two stars are 5 magnitudes apart (as in the example of the 1st magnitude star and the 6th magnitude star), they differ by a factor of 2.512^5 (2.512 to the fifth power) and a good pocket calculator gives the answer as 100. If they are 6 magnitudes apart, one is about 250

times brighter than the other. And if you're comparing, say, a 1st magnitude star with an 11th magnitude star, they are 2.512^{10} different in brightness, meaning a factor of 100 squared, or 10,000.

The faintest object visible with the Hubble Space Telescope is about 25 magnitudes fainter than the faintest star seen with the naked eye (assuming normal vision and viewing skills — some experts and a certain number of liars and braggarts say that they can see 7th magnitude stars). Twenty-five magnitudes is 5 times 5 magnitudes, meaning that it's a factor of 100^5 power fainter. So the Hubble can see 100 x 100 x 100 x 100 x 100, or 10 billion times fainter than the human eye. We expect nothing less from a billion-dollar telescope. At least it didn't cost $10 billion.

You can get a good telescope for under $1,000, and you can download the billion-dollar Hubble's best photos from the Internet for free at www.stsci.edu.

Don't make light of a light-year

The distances to the stars and other objects beyond the planets of our solar system are measured in *light-years*. As a measurement of actual length, a light-year is about 5.9 trillion miles long.

People confuse a light-year with a length of time, because the term contains the word *year*. But a light-year is really a distance measurement — the length that light travels, swooping through space at 186,000 miles per second, over the course of a year.

When people view an object in space, they see it as it appeared when the light left the object. Consider these examples:

✔ When astronomers spot an explosion on the Sun, we're not seeing it in real time; the light from the explosion takes eight minutes to get to Earth.

✔ The nearest star beyond the Sun, Proxima Centauri, is about 4 light-years away. We can never see Proxima as it is now; only as it was 4 years ago.

✔ Look up at the Andromeda Galaxy on a clear, dark night in the fall. It's the most distant object that you can readily see with the unaided eye. The light your eye is receiving left that galaxy about 2 million years ago. If the galaxy disappeared by some mysterious means, the people on Earth wouldn't even know for another two million years.

Here's the bottom line:

✔ When we look out into space, we are looking back in time.

✔ There's *no way* to know exactly what an object out in space looks like right now.

When we look at some big bright stars in a faraway galaxy, it's entirely possible, and often likely, that those particular stars don't even exist any more. Some massive stars only live for 10 or 20 million years. If we see them in a galaxy that's 50 million light years away, we're looking at lame duck stars. They aren't shining in that galaxy any more; they're dead.

If we sent a flash of light toward one of the most distant galaxies found with Hubble and other major telescopes, the light would take about 10 to 14 billion years to get there, because those galaxies are 10 to 14 billion light years away. The Sun, however, will swell up and destroy all life on Earth a mere 5 or 6 billion years from now. So the light would be a futile advertisement of our civilization's existence, a flash in the celestial pan.

Hey, you! A.U.

The Earth is about 93,000,000 miles from the Sun, or one Astronomical Unit (A.U.). The distances between objects in the solar system are usually given in A.U. Its plural is also "A.U." (don't confuse A.U. with "Hey you," which is what you say when you don't know somebody's name).

In public announcements, press releases, and popular books, astronomers state how far the stars and galaxies that they are studying are "from Earth." But among themselves and in technical journals, the distances are always given from the Sun, the center of the solar system. This rarely matters, because the distances of the stars are not measured precisely enough that one A.U. more or less would make a difference, but it's done this way for consistency.

The fixed stars are moving all the time

The stars used to be called the "fixed stars" to distinguish them from the wandering planets. But, in fact, the stars are in constant motion, both real and apparent. The whole sky rotates overhead because the Earth is turning. The stars rise and the stars set, like the Sun and the Moon, but they stay in formation. The stars in the Great Bear don't swing over to the Little Dog or Aquarius, the Water Bearer. Different constellations rise at different times, on different dates as visible from different places on Earth.

Actually, the stars in Ursa Major (and every other constellation) do move with respect to one another, at breathtaking speeds, measured in the hundreds of miles per second. But they're so far away that scientists need precise measurements over considerable intervals of time to detect their motions across the sky. So, 20 thousand years from now, the stars in Ursa Major will form a different pattern on the sky. Maybe it will even look like a Great Bear.

In the meantime, the positions of millions of stars have been measured, and lots of them are tabulated in catalogs and marked on star maps. They are listed in astronomers' units, called right ascension and declination, known to all astronomers, amateur and pro, as *RA* and *Dec.*

  ✔ The RA is the position of a star measured in the east-west direction on the sky (like longitude, the position of a place on Earth measured east or west of the prime meridian at Greenwich, England).

  ✔ The Dec is the position of the star measured in the north-south direction, like the latitude of a city, which is measured north or south of Earth's equator.

RA is usually listed in units of hours, minutes, and seconds, like time.

If you like using RA and Dec

A star at RA 2ʰ 00ᵐ 00ˢ is two hours east of a star at RA 0ʰ 00ᵐ 00ˢ, regardless of their declinations. RA increases from west to east, starting from RA 0ʰ 00ᵐ 00ˢ, which corresponds to a line in the sky (actually, half a circle, centered on the center of the Earth) from the North Celestial Pole to the South Celestial Pole. The first star might be at Dec 30° North and the second star might be at Dec 15° 25' 12" South, but they are still 2 hours apart in the east-west direction. (And 45° 25' 12" apart in the north-south direction.)

There are rules about the units of RA and Dec:

✔ An hour of RA equals an arc of 15 degrees on the equator in the sky. There are 24 hours around the sky. 24 x 15 = 360 degrees, or a complete circle around the sky. A minute of RA, called *a minute of time* is a measure of

angle on the sky that is ⅟₆₀th of an hour of RA So it's 15° ÷ 60, or ¼°. A second of RA, or a *second of time* is sixty times smaller than a minute of time.

✔ Dec is measured in degrees, like the degrees in a circle, and in *minutes and seconds of arc.* A whole degree is about twice the apparent or angular size of the full Moon. Each degree is divided into 60 minutes of arc. There are about 32 minutes of arc (32') across the Sun or the full Moon. And each minute of arc is divided into 60 seconds of arc (60"). When you look through a backyard telescope at high magnification, the image of a star is blurred by turbulence in the air. Under good conditions (low turbulence), the image is about 1" or 2" across.

A few simple rules may help you remember how RA and Dec work and how to read a star map (see Figure 1-3):

✔ The North Celestial Pole (NCP) is the place on the sky to which the axis of Earth points in the north direction. If you are standing at the geographic north pole, the NCP is right overhead.

✔ The South Celestial Pole (SCP) is the place on the sky to which the axis of Earth points in the south direction. If you are standing at the geographic south pole, the NCP is right overhead. I hope you dressed warmly, you're in Antarctica!

✔ The imaginary lines of equal RA run through the NCP and SCP, and they are really semicircles that are centered on the center of Earth. They may be imaginary, but they are marked on most sky maps, to help people find the stars at particular RAs.

✔ The imaginary lines of equal Dec, like the line in the sky that marks Dec of 30° North, pass right overhead at the corresponding geographic latitudes. So if you are standing in New York City, latitude 41° North, the point right overhead is always at Dec 41° North, although its RA changes constantly as Earth turns. These imaginary lines are on star maps, too, and they are called *declination circles*.

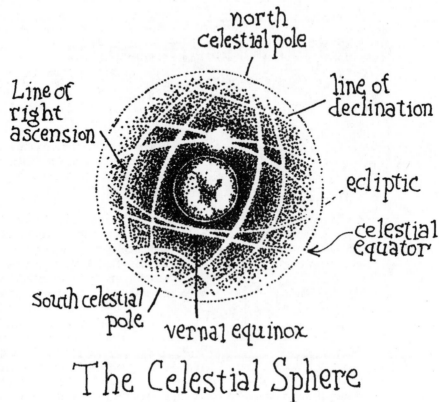

north
celestial pole

line of
declination

Line of
right
ascension

ecliptic

celestial
equator

South celestial
pole

vernal equinox

The Celestial Sphere

Figure 1-3:
Decoding
the celestial
sphere.

Do astronomy at home

If you have a patch of yard with a view of the sky — not too many trees and neighboring homes blocking the horizon — you're all set. You can set up a telescope on a clear night, or take out a pair of binoculars, and start spotting the stars. If you live in the middle of downtown Los Angeles, where the sky is white with stray light from the city, you can join an astronomy club that travels to a darker viewing location. Often, they just head for the hills.

And if you're mostly interested in science — the discoveries astronomers are making — you can catch the news in monthly magazines designed especially for amateur astronomers. Even better, you can visit free Web sites that tell you everything you ever wanted to know about what's going on in the sky, plus more you didn't know enough about to ask.

Astronomy is ideal for families. Set up your telescope and everyone lines up for a peek. Can't get a babysitter? Bundle up the kids and bring them along to the star party. They'll even help carry the telescope. Bring blankets and sleeping bags. What better way to think about the world than to gaze at the spectacle of starry heaven as you slowly fall asleep?

Suppose that you want to find the NCP as visible from your backyard. Face due north, and look at an altitude of X degrees, where X is your geographic latitude. (I'm assuming that you live in North America, Europe, or someplace else in the Northern Hemisphere. If you live in South America, South Africa, Australia, or any place in the Southern Hemisphere, you can't see the NCP. Instead you should look for the SCP. It's the spot that's due south and whose altitude in the sky, measured in degrees above the horizon, is equal to your geographic latitude.)

In almost every astronomy book, the symbol " means seconds of arc, not inches. But at every university, there's always a student in Astronomy 101 who will write on an exam, "The image of the star was about 1 inch in diameter." Understanding beats memorizing every day, but not everyone understands.

Here's the good news: If you just want to spot the constellations and the planets, you don't have to find out how to use RA and Dec. You just have to compare a star map prepared for the right time of year and time of night (as printed, for example, in many daily newspapers) with the stars that you see at about that date and time of night. But if you want to understand how star catalogs and maps work, and how to home in on faint galaxies with your telescope, understanding the system helps.

And if you get one of those snazzy, new, surprisingly affordable telescopes with computer control (see Chapter 3), you can punch in the RA and Dec of a comet that's just been discovered and announced, and the 'scope will point right at it. (Every comet announcement is accompanied by a little table or *ephemeris* of the predicted RA and Dec of the comet, night by night, as it sweeps across the sky.)

Gravity: A Force to be Reckoned With

Ever since Newton (Isaac, not the cookie), everything in astronomy has revolved around gravity. Newton explained it as a force between any two objects. The force depends on mass and separation. The more massive the object, the more powerful its pull. The greater the distance, the weaker the gravitational attraction.

Einstein developed an improved theory of gravity, which passes experimental tests that old Isaac's theory flunks. Newton's theory was good enough for commonly experienced gravity, like the force that made the apple fall on his head (if it really did hit him). But Einstein's theory also predicts effects that happen close to massive objects, where gravity is very, very strong. To Einstein, gravity wasn't really a force, it was the bending of space and time by the very presence of a massive object, such as a star. I get all bent out of shape, just thinking about it.

Newton's gravity explains the following:

- ✔ Why the Moon orbits Earth, Earth orbits the Sun, the Sun orbits the center of the Milky Way, and lots of other orbits, too
- ✔ Why a star or a planet is round
- ✔ Why gas and dust in space start clumping to form new stars

Einstein's theory of gravity, the General Theory of Relativity, explains the following:

- ✔ Why stars visible near the Sun during a total eclipse seem slightly out of position
- ✔ Why black holes may exist
- ✔ Why as the Earth turns, it drags warped space and time around with it, an effect that some scientists say they have measured, while others are waiting for more definitive observations

You can find out about black holes in Chapter 10, but you won't have to know the General Theory of Relativity. You can get smarter if you read every chapter. But your friends won't call you Einstein unless you let your hair grow, parade around in a messy old sweater, and stick out your tongue when they take your picture.

But relativity theory is important to how scientists study the universe today. Knowing that "everything is relative" and appreciating the paradoxical nature of much of the universe (yes, light *is* both a particle and a wave) have opened the lid to a Pandora's treasury of astronomical speculation and progress.

A Commotion of Motion

Everything in space is moving and turning. Objects can't sit still. Another body is always pulling on any star, planet, galaxy, or spacecraft. The universe has no center.

For example, Earth is

- ✔ Turning on its axis, what astronomers call *rotating*. Earth takes one day to turn all the way around.
- ✔ Orbiting around the Sun, what astronomers sometimes call *revolving*. One complete orbit takes one year.
- ✔ Traveling with the Sun in a huge orbit around the center of the Milky Way, taking about 226 million years to go around once. The duration of that trip is called the *galactic year*.

✔ Moving with the Milky Way in a trajectory around the center of mass of the *Local Group of Galaxies,* a couple dozen galaxies in our neck of the universe.

✔ Moving through the universe with the Local Group as part of the *Hubble Flow,* the general expansion of space caused by the Big Bang.

And you, as a person on Earth, are participating in the motions of rotation, revolution, galactic orbiting, Local Group cruising, and cosmic expansion. You do all that while you're driving to work, and maybe you didn't even know it. Ask for a little consideration the next time you are a few minutes late.

Remember Ginger Rogers? She did everything Fred Astaire did when they danced in the movies, and she did it all backward. Like Ginger and Fred, the Moon follows all the motions of Earth (though not backward) except for Earth's rotation; the Moon rotates more slowly, about once a month. And, it does all this while also revolving around Earth (which it also does about once a month).

The Big Bang is the theoretical event that gave rise to the universe and set space itself expanding at a furious rate. It explains lots of observed phenomena, and it predicted some that hadn't been observed before the theory was promulgated. It's the best beginning-of-the-universe theory that astronomers have.

Well, this has to be a great book. Even the first chapter ends with a Bang!

Chapter 2

Skywatching: Join the Crowd

● ●

In This Chapter

▶ Joining astronomy clubs and using other resources

▶ Visiting observatories and planetariums

▶ Enjoying star parties, eclipse tours, and telescope motels

● ●

A stronomy has universal appeal. The stars have fascinated people everywhere from cave times. Early observations of the sky led to all sorts of theories about the universe and attributions of power and purpose to the movements of stars, planets, and comets. As you look up at the sky, thousands if not millions of people worldwide are watching with you. And over time viewers like these have provided the basis for the modern understanding of the sky and its inhabitants. When it comes to skywatching, you are not alone. You have many people and publications and other resources to get you started, keep you going, and help you participate in the great work of explaining the universe.

In this chapter, I introduce you to these resources, and suggest how you can get started. The rest is up to you. So join in!

All Eyes on the Skies: You Are Not Alone!

Lots of readily available information, organizations, people, and facilities can help you get started and keep going in astronomy. You can find Web sites with basic information on astronomy and current events in the sky. You can join associations and activities to help researchers keep track of stars and planets, and you can attend astronomy club meetings, lectures, and instructional sessions, which allow you to share telescopes and viewing sites, and enjoy the sky with others.

Join an astronomy club for star-studded company

The best way to break into astronomy without undue effort and expense is to join an astronomy club and meet the regulars. They hold monthly meetings where the old hands pass on tips on techniques and equipment to beginners and where local and visiting scientists present talks and slide shows. Members of these clubs already know where to get a good deal on a used telescope or binoculars, and which products on the market are worth the money.

Even better, the clubs sponsor observing meetings, usually on weekend nights and occasionally on special dates when a meteor shower, an eclipse, or another special event occurs. You can find out more about astronomy practice and equipment at an observing meeting than anyplace else. You don't even need to bring a telescope; most folks are happy to give you a look through theirs. Just dress warmly, wear sensible shoes, bring mittens, and put on a smile!

If you live in a city or a close suburb, chances are good that your night sky is bright, and you'll do better if you observe at a dark spot in the country. Your local astronomy club has probably already found a good spot, and when they converge on that lonely place, there's safety in numbers. So join up!

If you live in a good-sized city or a college town, chances are that an astronomy club is active near you. To find the nearest astronomy club, consult the Web site of the U.S. "club of clubs," the Astronomical League, at www.astroleague.org. Browse through the list of over 200 member

---

U.S., Canadian, and other clubs

The Astronomical Society of the Pacific, with headquarters in San Francisco, publishes *Mercury* magazine for amateurs. It holds an annual meeting that moves around the western United States, and sometimes comes as far east as Boston or Toronto. Their Web site is www.aspsky.org. They offer numerous educational materials in astronomy to school teachers, as well.

Do you live in Canada? The Royal Astronomical Society of Canada has 23 Centres, which is a fancy name for astronomy clubs. Often, a professional or two from the nearest university is involved with the Centre's activities. You can find a list of the Centres on the RASC Web site, www.rasc.ca.

In the United Kingdom, the organization of choice is the British Astronomical Association, founded over a century ago. Their Web site is www.ast.cam.ac.uk/~baa.

Most other countries have astronomy clubs, too. Astronomy is truly a "universal" passion.

societies, arranged by state. For each one, the site provides a contact person's name, phone number, and Internet address. The Astronomy Mall has a Web site with club listings around the world at `astronomy-mall.com/regular/clubs-etc/clubsetc.html`. And Claude Marcotte has a wonderful list at `www.dsuper.net/~leia/clubs/usa.html`. Some local clubs have their own Web sites, too.

Check out resources: Web sites, magazines, and more

Finding out about astronomy is easy. You can choose from a wide range of resources, including Web sites, magazines, and some innovative new software. The following sections offer some tips for finding the best information.

Cyberspace

The World Wide Web offers sites on every topic in astronomy, and they are increasing at a, well, astronomical rate! One of the best is maintained by the editors of *Sky & Telescope* magazine, at `www.skypub.com`.

Get started observing the sky by checking out *Sky & Telescope*'s "This Week's Sky at a Glance" page, at `www.skypub.com/sights/sights.shtml`. Each week, it presents a color picture showing how the most interesting part of the sky appears to the naked eye. The Moon and bright planets are indicated and named, along with a direction on the horizon. Just face the indicated direction and spot the Moon, for example. Then you can pick out the planets and bright stars in relation to the Moon. The sky changes every day, so you can find daily summaries as well. Some nights, the Moon isn't visible, and you have to work a little harder to recognize objects and patterns in the sky.

To plan your viewing a month at a time, consult the monthly summary of sky events at *Astronomy* magazine's site, `www2.astronomy.com/astro`. If a meteor shower is coming up, you may want to plan a trip to a safe, dark viewing location in the country.

You can find pointers to these and other Web sites throughout this book; if you want more information on planets, comets, meteors, or eclipses, the Web offers good sites on every topic.

Completely puzzled? At some Web sites, NASA scientists will answer your questions. You can "Ask a NASA Scientist" at `imagine.gsfc.nasa.gov/docs/ask_astro/ask_an_astronomer.html` and if they don't know the answer, try "Ask the Space Scientist" at `image.gsfc.nasa.gov/poetry/ask/askmag.html`.

Publications

You can purchase excellent magazines to expand your knowledge of astronomy and your skill at practicing it. Most amateur astronomers subscribe to at least one of them. And, in many cases, if you join a local astronomy club, a subscription to a national magazine may be available at a member's discount.

Which astronomy magazine is right for you? That's like asking which car you should drive or which kind of shoes to buy. You probably choose a car after test-driving some. And, if the shoes fit, you buy them. Use the same test for a magazine: Head for the nearest big bookstore and check the magazine racks. Or, if your town has a shop that specializes in magazines and out-of-town newspapers, that's the place to go. Planetariums and science museums usually carry these magazines, too.

Pick up a copy of each of the "Big Two" astronomical magazines: *Sky & Telescope* and *Astronomy*. Test drive them for a month, and if you get a lot more out of one than the other, go ahead and subscribe.

Canadian readers can get the bimonthly *SkyNews,* a slick, full-color "Canadian Magazine of Astronomy & Stargazing," available from the National Museum of Science and Technology Corporation. Call 800-267-3999 for information.

In France, the magazine is Ciel & Espace; in Australia, it's *Sky & Space;* wherever you live, there's an astronomy magazine just for you.

And wherever you live, you need the annual *Observer's Handbook* of the Royal Astronomical Society of Canada (www.rasc.ca), compiled by over three dozen experts to help you enjoy the skies.

Software

A planetarium program for your personal computer is a real plus. The program can show you what the sky looks like from your home every day. This software is terrific to look at before you step outside to view the night sky. Some astronomers use these programs to plan their observing sessions. They prepare schedules of objects to scan with telescopes and binoculars at different times of the night, so they can use their "dark time" effectively.

Personal planetarium programs are available over a wide price range with a number of different features. Look for the newest ones that are nationally advertised in astronomy and science magazines and on Web sites; these tend to be the best because they take advantage of new technology and research. You need only one program to get started, however, and that may be the only program you'll ever need.

Here are two programs that I find useful:

✔ Although I've paid $300 or more for planetarium programs, I've recently enjoyed using one that retails for around $90 (but I found it on sale for $60). It's *Starry Night Deluxe,* from Sienna Software in Toronto. Sienna's Web site is www.siennasoft.com. You need a CD-ROM drive on your computer to run this and most other late-model planetarium programs. After you've installed the program, just click on the Starry Night icon and up comes a color picture of the sky, complete with a horizon and a few trees.

If you access the program at night, the picture shows how the stars, planets, and Moon look to the naked eye outdoors at your location (assuming the weather is clear). If you run the program during the day, you get a picture of the blue sky, with the Sun at the proper altitude. Turn on the program shortly after sunset, and minute by minute, you see the sky darken and the planets and stars come out. It's just like being outside, but you haven't left your computer yet! There are no red sails in this sunset; the program's scientific, not romantic. Click on icons, drag objects in the computer sky with your mouse, or change some settings and you can see what the sky will look like at any place on Earth at almost any time.

✔ *TheSky* is a highly regarded planetarium program with lots of bells and whistles. It's available in several versions, for beginners and advanced astronomers, at prices from $129 to $249. Produced by Software Bisque in Golden, Colorado, you can check it out on their Web site, www.bisque.com/thesky.

The best way to select the planetarium program that's right for you is to talk to experienced amateur astronomers at your local astronomy club. What works for them should work for you.

Visit Observatories and Planetariums for Guided Viewing

You can visit professional observatories (organizations that have large telescopes staffed by astronomers and other scientists for use in studying the universe) and public planetariums (specially equipped facilities with machines that project stars and other sky objects in a darkened room with plain English explanations of various sky phenomena) and find out more about telescopes, astronomy, and research programs.

Observatories

You can find dozens of professional observatories in the United States and many more abroad. They are research institutions, operated by colleges and universities, or by government agencies. They include the U.S. Naval Observatory (located in the heart of Washington, D.C. and guarded by the Secret Service — the Vice President lives on the grounds) and outposts on remote mountaintops (such as Denver University's Mt. Evans Meyer-Womble Observatory, billed as the "Highest Operating Observatory on Earth, at 14,148 feet elevation"). Some observatories are dedicated entirely to public education and information, often operated by cities, counties, school systems, or nonprofit organizations.

The research observatories are usually located in exotic or aptly named places, such as the Lowell Observatory on Mars Hill, in Flagstaff, Arizona, where the planet Pluto was discovered in 1930. Founder Percival Lowell thought he could see canals on Mars through a telescope there.

The National Solar Observatory runs a cluster of sun-watching telescopes at Sunspot, New Mexico, high above Cloudcroft, which is high above Alamogordo. And Georgia State University in Atlanta operates an observatory out in the country at Hard Labor Creek. (Don't trespass there!)

Many observatories have an open house for the public at weekly or monthly intervals, and some even offer daily tours (during the day) and operate astronomical museums, complete with souvenir shops.

Some observatories are even located in or near busy cities, such as the Griffith Observatory and Planetarium, operated entirely for the public at Griffith Park in Los Angeles, and the Mount Wilson Observatory in the San Bernadino mountains above Los Angeles, California. Mount Wilson, where the expansion of the universe and the magnetism of the Sun were discovered, is well worth a visit.

At Palomar Observatory, near San Diego, California, you can see the famous 200-inch telescope, which for decades was the largest and best in the world. With new instrumentation, the telescope is still a great contributor of new knowledge on the universe.

One of the largest observatories in the continental United States encourages visitors. It's the Kitt Peak National Observatory, 56 miles west of Tucson, Arizona. The National Optical Astronomy Observatories runs it. A whole bunch of the largest and most advanced telescopes of the United States, Canada, Japan, and the United Kingdom — known collectively as the Mauna Kea Observatories — can be found atop the Mauna Kea volcano on the island of Hawaii.

You can also visit radio astronomy observatories, where scientists "listen" to radio signals from the stars, or even seek signals from alien civilizations. The National Radio Astronomy Observatory, for example, has facilities near

Socorro, New Mexico, and Green Bank, West Virginia, that you can visit, and another on Kitt Peak.

You can find lists of observatories on both the *Astronomy* and *Sky & Telescope* magazine Web sites. Call the observatory near you or send an e-mail and find out when you can visit. If you find the observatory really interesting, go again and again. The tour is usually free, but a contribution box or jar is often available. Hang around enough, show some interest, and you too might become an unpaid intern in an exciting research project.

In the early '60s, I used to give tours at an observatory in the Midwest. Some nice folks even tipped me a quarter now and then. When you're a grad student making $1,200 a year, 25 cents is nothing to sneeze at.

Here are the Web sites of observatories in this section, in the order in which I mention them; some offer visiting hours and public tours, and some don't:

- U.S. Naval Observatory www.usno.navy.mil
- Mt. Evans Meyer-Womble Observatory www.du.edu/physastron
- Lowell Observatory www.lowell.edu
- National Solar Observatory www.nso.noao.edu/welcome.html
- Georgia State University www.chara.gsu.edu/HLCO/hlco.html
- Griffith Observatory and Planetarium www.griffithobs.org
- Mount Wilson Observatory www.mtwilson.edu
- Palomar Observatory astro.caltech.edu/palomarpublic/index.html
- Kitt Peak National Observatory www.noao.edu/kpno/kpno.html
- National Optical Astronomy Observatories www.noao.edu
- Mauna Kea Observatories www.ifa.hawaii.edu/ifa/observatories.html
- National Radio Astronomy Observatory www.nrao.edu

Planetariums

Planetariums, also called *planetaria,* are just right for a beginning astronomer. They provide instructive exhibits and wonderful sky shows, projected indoors on the planetarium dome or a huge screen. And many offer nighttime skywatching sessions with small telescopes, usually outside in the parking lot or at a nearby public park. Many of them have excellent shops where you can browse through the latest astronomy books, magazines, and star charts. The planetarium staff can direct you to the nearest astronomy club. The club may even meet after hours in the planetarium itself.

I practically grew up in the Hayden Planetarium at the American Museum of Natural History in New York City. Occasionally, I confess, I even snuck in for free. They were nice enough to have me back to speak (also for free) at their 50th anniversary. Although my old planetarium in New York has been torn down, they are putting up a spectacular new one that should be a prime destination next time you visit the Big Apple. It will be a lot cheaper than a Broadway show, and its stars will never miss a cue or sing off key.

Loch Ness Productions, in Massachusetts, is the monster of the planetarium business. They keep tabs on almost 3,000 planetariums worldwide. To find the planetarium near you or a location you may visit, check their Web site at `www.lochness.com`.

Vacation with the Stars: Star Parties, Eclipse Cruises, and Telescope Motels

An astronomy vacation is a treat for the mind and a feast for the eyes. Plus, it's often cheaper than a conventional holiday. You don't have to visit the hottest tourist destinations to keep up with the Joneses. You can have the experience of a lifetime and come back raving about what you saw and did, not just what you ate and spent.

It *is* possible, however, to blow big bucks on one type of astronomy vacation: the eclipse cruise. But if you like ocean cruises, taking one to an eclipse doesn't cost any more than similar voyages that have no celestial rewards. And bargain-basement eclipse tours are available, too.

Star parties

Star parties are outdoor conventions of amateur astronomers. Hundreds of telescopes are set up in a field and people take turns looking through them. (Be prepared to hear lots of "Oohs" and "Ahs.") Prizes are given for the best homemade telescopes and equipment. If it rains in the evening, party goers may watch slide shows in a nearby hall or a big tent. Arrangements vary, but often some attendees camp in the field, while others rent inexpensive cabins or commute from nearby motels. PBS television's *Mysteries of Deep Space* series, which is rerun at intervals, followed one family as they packed up a van and headed cross country for the Texas Star Party, one of the most famous of its type.

Star parties usually last several days and nights (sometimes as much as a week). They attract a few hundred to a few thousand (yes, thousand!) telescope makers and amateur astronomers. And the larger star parties have Web sites with photos of previous events and details on coming attractions.

The leading star parties in the United States include

- ✔ Stellafane, in Vermont (`www.stellafane.com`)
- ✔ Texas Star Party, where you can see thousands of stars from a remote location in the Lone Star State (`www.metronet.com/~tsp`)
- ✔ Riverside Amateur Telescope Makers, for California dreamers fascinated by truly *celestial* bodies (`www.rtmc-inc.org`)
- ✔ Enchanted Skies Star Party, out on the desert in New Mexico (`www.socorro-nm.com/starparty.html`)
- ✔ Nebraska Star Party, which claims the darkest site in the continental United States (`www.4w.com/nsp`)

Eclipse cruises and tours: On the path of totality

Eclipse cruises and tours are planned voyages to the places where total eclipses of the Sun are visible. Astronomers can calculate long in advance when and where an eclipse will be visible. The locations where you can see the total eclipse are limited to a narrow strip across land and sea, the *path of totality.* You can stay home and wait for a total eclipse to come to you, but chances are, you won't live long enough to see more than one, if any. Or, you can travel to the path of totality. The next total eclipse of the Sun visible from the continental United States is not until 2017.

Booking a tour

If an eclipse is within easy driving distance, you don't need to sign up for a tour. And if you're an experienced domestic and international traveler, you could arrange to go on your own to the path of totality of a distant eclipse. But consider this fact: Expert meteorologists and astronomers identify the best viewing locations years in advance. More often than not, these places are not vast metropolises that offer huge numbers of vacant accommodations. They are random spots on the globe. Once a spot is recognized as a prime location for a coming eclipse, tour promoters and savvy individuals book all or most of the local hotels and other facilities years in advance. Johnnies-or-Janies-come-lately, especially those traveling on their own, may be out of luck.

The tour promoter usually engages a meteorologist and a few professional astronomers (sometimes even me). So you have the benefit of a weatherperson to make last-minute decisions on moving the group's observing site to a place with a better next-day forecast, an astronomer to show you the safest methods to photograph the eclipse, and usually another lecturer who tells old eclipse tales and reports on the latest discoveries about the Sun and space.

And, if you're out in the boondocks, at least one of the experts carries a Global Positioning System receiver, so signals from satellites can be used to verify that the group is at the desired location.

On the night following the eclipse, everyone shows their videotapes of the darkening sky, the birds coming to roost, a klutz knocking over his telescope at the worst possible time, and the excited crowd saying "Wow" and "Hurray." And of course, they replay the eclipse — over and over.

If these reasoned arguments have not convinced you to pick an eclipse tour when you want to see a solar eclipse, consider this: A group tour to a foreign destination is almost always cheaper than going on your own.

Cruising is better

An eclipse cruise is better than a tour, but more expensive. It's better because at sea, the captain and navigator have "two degrees of freedom." When the meteorologist says "head southwest down the path of totality for 200 miles" on the night before the eclipse (because that's his or her best forecast for a cloud-free location at eclipse time), the ship can follow those instructions. But on land, you have to keep the bus on the road, and there may not be a road going the direction you need to go. If a road is available, it may be jammed by thousands of eclipse-trippers trying to get a better site at the last minute because the weather changed. On the cruise, leave the steering to the crew, recline in your deck chair, sip a piña colada, load your camera, and wait for totality.

I've viewed many eclipses and my experience is that if you stay on the ground, you get to see totality about half the time and are clouded out half the time. But if you're on an ocean liner, you never miss. (Well, there's always a first time.)

Making the decision

Eclipse tours and cruises are advertised in astronomy magazines and in many science and nature magazines, and are available through travel agents. The cruises are often sponsored by clubs, fraternal organizations, and alumni societies.

Here's a way you can choose the right tour or cruise for you:

✔ Consult current and back issues of astronomy magazines. Most run articles about the viewing prospects for a solar eclipse a few years in advance. Get their expert recommendations about the best viewing sites.

✔ Check out the advertisements from travel operators. Which tours and cruises go to the best places? Get brochures from travel agents, tour promoters, and cruise lines. Often, promoters list previous successful eclipse trips, indicating that they are experienced.

EYES ON THE SKIES

Scientific research you can participate in

Your astronomy hobby can be beneficial as well as fun. You can contribute to the march of science by joining national and worldwide efforts to gather precious scientific data. Let's face it, you may have only a pair of binoculars, while the Keck Observatory in Hawaii has two 10-meter (400-inch wide) telescopes. But if it's cloudy on Mauna Kea, Keck will see nothing. And if a spectacular fireball shoots over your hometown, you may see it although no professional expert does.

One of the most spectacular and interesting meteors of all time was recorded by secret Department of Defense satellites, and by an amateur movie maker vacationing at Glacier Lake National Park. A clip from that home movie appears in just about every scientific documentary about meteors, asteroids, and comets that appears on TV. It pays to be in the right place at the right time. And some day, you will.

Join other amateur astronomers and enjoy the projects I recommend throughout this book.

You can do these activities on your own, but it's always easier to compare notes with someone who's experienced, so ask around at the local astronomy club.

You can find a list of astronomical travel outfits on the *Sky & Telescope* Web site at www.skypub.com/resources/marketplace/travel.html. Also, check out the "Marketplace" section of *Astronomy* magazine.

Telescope motels

Telescope motels are resorts where the attractions are the dark skies and the opportunity to set up your own telescope in an excellent viewing location. They usually have telescopes of their own that you can use, perhaps at an additional fee.

There are at least nine telescope motels worldwide. Three of the best known ones in the United States are

- Star Hill Inn, at altitude 7,200 feet in Sapello, New Mexico, a pioneer in resort lodgings for astronomers (www.starhillinn.com)
- Skywatcher's Inn, in Benson, Arizona, which features a 20-inch telescope (www.communiverse.com/skywatcher)
- Molokai Ranch in Hawaii, where you can see more of the southern stars than anywhere else in the United States (www.molokai-ranch.com)

Molokai Ranch's Web site emphasizes kayaking, biking, horseback riding, and more, and lists annual stargazing events. But you can check out the stars from Molokai whether there's a special event or not.

In recent years, most new, large observatories in the United States have been erected in Arizona, Hawaii, and New Mexico. The skies that are darkest and finest for the professionals are there for you to enjoy, too, at the telescope motel of your choice!

Once you know the resources, organizations, facilities, and equipment that can help you enjoy astronomy more deeply, you can move comfortably on to the science of astronomy itself, the nature of the objects and phenomena out there in deep space.

I describe the equipment you need in order to get started in Chapter 3, so read on!

Chapter 3

The Way You Watch Tonight: Observing the Skies

In This Chapter

▶ Observing the skies with your naked eyes

▶ Finding objects

▶ Choosing equipment for astronomy

▶ Understanding observational fundamentals

*I*f you have ever gone outside and looked at the night sky, you've been stargazing — observing the stars and other objects in the night sky. Naked-eye observation can distinguish colors and the relationships between objects — like finding the North Star using the "pointer stars" in the bowl of the Big Dipper.

From naked-eye observation, it's a short step to adding optics to see fainter stars and to view objects with greater detail. First try binoculars, and then graduate to a telescope. Next thing you know, you're an astronomer!

But I'm getting ahead of myself. First, you need to take some quiet looks at the cosmos and see for yourself its beauty and mystery. You can use three basic tools — at least one of which you already own.

Whether you use your own eyes, a pair of binoculars, or a telescope, each method of observation is best for some purposes.

✔ The human eye is ideal for watching meteors or the aurora borealis, or enjoying a conjunction of the planets (when two or more planets are close to each other in the sky) or of a planet and the Moon, for example.

✔ Binoculars are best for observing bright variable stars, which are too far from their comparison stars (stars of known constant brightness used as a reference to estimate the brightness of a star that varies in brightness) to be seen together through a telescope. And binoculars are wonderful for sweeping through the Milky Way and viewing the bright nebulae and star clusters that dot it here and there. Some of the brighter galaxies —

> such as M31 in Andromeda, the Magellanic Clouds, and M33 in Triangulum — look best through binoculars also.
>
> ✔ A telescope is needed to get a decent look at most galaxies, and to distinguish the members of close binary stars, among many other uses.

Begin with Naked-Eye Observation

The most important step in observing with the naked eye is to shield your vision from interfering lights. If you can't get to a dark place in the country, at least find a dark spot in your backyard or possibly on the roof of your building. You won't eliminate the light pollution high up in the sky that results from the collective lights of your city, but trees or the wall of a house might prevent nearby streetlamps from shining in your eyes.

Watching the bright Comet Hyakutake in 1996 from a small city in the Finger Lakes of northern New York State, I found that just walking around the corner of a building, so that I was in the building's shadow rather than exposed to lights on the adjacent main road, made an enormous difference in the visibility of the comet.

If you don't already know the geographical directions in your area, take the time to learn them. Then you can use the weekly sky highlights from the *Sky & Telescope* Web site, or the picture that fills your screen when you run a desktop planetarium program, to orient yourself to the brightest stars and planets. When you recognize the bright stars, you can more easily pick out the patterns of slightly fainter ones all around them.

Table 3-1 lists some of the brightest stars you can see in the night sky, along with the constellations you find them in. Many of them are visible from the continental United States and Canada. A few are only readily seen from southern latitudes. So the bright stars not visible to United States readers may be prominent sights for Australians. See Chapter 11 for information on spectral class.

Start your observations by consulting a star map or desktop planetarium; then see how many of these stars you can locate at night. Next, identify some of the dimmer stars in the same constellations. And, of course, keep your eye out for the bright planets: Mercury, Venus, Mars, Jupiter, and Saturn.

Table 3-1	Brightest Stars as Seen from Earth		
Common Name	*Apparent Magnitude*	*Constellation Designation*	*Spectral Class*
Sirius	−1.5	α Canis Majoris	A
Canopus	−0.7	α Carinae	A
Rigil Kentaurus	−0.3	α Centauri	G
Arcturus	−0.04	α Bootis	K
Vega	0.0	α Lyrae	A
Capella	0.1	α Aurigae	G
Rigel	0.1	β Orionis	B
Procyon	0.4	α Canis Minoris	F
Achernar	0.5	α Eridani	B
Betelgeuse	0.5	α Orionis	M
Hadar	0.6	β Centauri	B
Acrux	0.7	α Crucis	B
Altair	0.8	α Aquilae	A
Aldebaran	0.9	α Tauri	K
Antares	1.0	α Scorpii	M
Spica	1.0	α Virginis	B
Pollux	1.1	β Geminorum	K
Fomalhaut	1.2	α Piscis Austrini	A
Deneb	1.3	α Cygni	A

In winter and summer, the Milky Way runs high in the sky at most locations in the United States. If you can recognize the Milky Way as a wide, faintly luminous band across the sky, you have at least a pretty fair observing site.

Ideally, you want a site that has a good horizon, with only trees and low buildings in the distance, but finding that kind of location is next to impossible in a major urban area.

If you fail to find a site with a good horizon in all directions, the most important horizon is the southern one. Most observations in the Northern Hemisphere of the Earth are made while facing roughly south, so that east is on your left and west is on your right. As you face south, the stars are rising to your left and setting to your right. If you live in the Southern Hemisphere or are visiting there, reverse the directions and face north.

Always have a watch, a notebook, and a dim or red flashlight to use for recording what you see. Some flashlights come with a red bulb, or you can buy red cellophane from a greeting card store to wrap around the lamp.

Seeing Stars: A Primer on Sky Geography

The Earth turns. That concept was proclaimed in the 4th century B.C. by the Greek philosopher Heraclides Ponticus. But people doubted Heraclides's observations because folks thought they should feel dizzy like riders of a fast merry-go-round or a whirling chariot. They couldn't imagine a turning Earth if they could not physically feel the effects. Instead, our ancient forebears thought the Sun was racing around the Earth, making a complete revolution every day.

Proof of Earth's turning, or *rotation,* didn't come about until 1815, more than two millennia after Heraclides (there wasn't much government funding for research back then, so progress was slow but it cost less). The proof came from a big French swinger: a heavy metal ball hanging from the ceiling and suspended above the floor of the Pantheon (a church) in Paris on a 200-foot wire. This ball is called a *Foucault pendulum,* after the French physicist who came up with the plan. If you kept an eye on the pendulum as it swung back and forth all day, you could see that the direction taken by the swinging ball across the floor gradually changed, as though the floor was turning underneath it. And it was; the floor was turning with the Earth.

If you're not convinced that the Earth turns, or you just like to watch swingers, you can see a Foucault pendulum in the National Museum of American History at the Smithsonian Institution in Washington, D.C. If you believe, however, you can just have a drink as you enjoy a sunset.

As the Earth turns . . .

As I explain in Chapter 1, the rotation of Earth around its axis makes the stars and other sky objects appear to move across the sky, from east to west. In addition, the Sun moves across the sky during the year on a circle called the *ecliptic*. The ecliptic is inclined by 23.5 degrees to the celestial equator, the same angle by which the axis of Earth is tilted from the perpendicular to its orbital plane.

The planets stay close to the ecliptic as they move throughout the year. There are twelve constellations arranged around the ecliptic, which collectively are called the *Zodiac:* Aries, Taurus, Gemini, Cancer, Leo, Virgo, Libra, Scorpius, Sagittarius, Capricorn, Aquarius, and Pisces. (Actually, a 13th constellation intersects the ecliptic — Ophiuchus — but in ancient times it was not included with the original 12.)

As Earth proceeds along its orbit around the Sun, the rising and setting times of the stars are earlier by about 4 minutes each night. This results in a different appearance of the night sky as Earth progresses through its seasons. The stars are not located in the same spots throughout the night or throughout a year. The constellations that were high in the sky at dusk a month ago are lower in the west at dusk now. And if you view the constellations that are low in the east just before dawn, it's a preview of what you'll see at midnight in a few months.

To help you keep track of all this coming and going (if you don't have a telescope that will do it for you), use the sky maps that come monthly in astronomy magazines such as *Sky & Telescope* or *Astronomy*. Your daily newspaper may also contain maps of the local skies. Or you can get an inexpensive *planisphere,* in which the night sky is represented by a rotating wheel in a frame with a hole cut into it that represents the limits of your view. It can be adjusted according to the time and date.

Finding the North Star (Polaris)

Of course, anyone can walk outside on a clear night and see some stars. But how do you know what you're seeing? How can you find it again? What should you watch for?

One of the most time-honored ways of getting familiar with the night sky if you live in the Northern Hemisphere is to become familiar with the North Star, or Polaris, which barely moves. Once you know north, you can orient yourself to the rest of the northern sky, or to anywhere else for that matter. In the southern sky, you need to find the bright stars Alpha and Beta Centauri, which point the way to the Southern Cross.

How bright is bright?

I talk about magnitude in Chapter 1, but you should know that magnitude can actually be given three different ways:

☛ *Absolute magnitude* is what scientists have determined is the true brightness of the sky object, as visible from a standard distance of 32.6 light-years.

☛ *Apparent magnitude* is how bright the object appears from Earth, which can be different from its absolute magnitude, depending on how far away from Earth the sky object is located. A star closer to Earth may appear brighter than one farther away, even if its absolute magnitude is fainter.

☛ *Limiting magnitude* is related to the condition of the viewable sky at the time of observation — how clear and dark the sky is. A very bright object may be invisible if atmospheric conditions get in the way. Limiting magnitude is used most often in meteor and deep sky observations. On a clear dark night, the limiting magnitude may be 6 at the zenith, but in the city the limiting magnitude may be only 4.

Star charts depict the apparent magnitudes of the stars, to simulate their appearances in the sky.

You can easily find the North Star using the Big Dipper in the constellation Ursa Major. The Big Dipper is one of the most easily recognized sky patterns (see Figure 3-1). If you live in the continental United States, you can see it every night of the year.

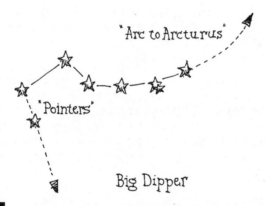

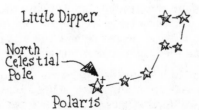

Figure 3-1:
The Big Dipper is a Big Pointer to other sky-sights.

The two brightest stars in the Big Dipper, Dubhe and Merak, form one end of its bowl and point directly to the North Star, or Polaris. Other stars that use the bowl of the Big Dipper as signposts include Castor and Pollux in Gemini and Deneb in Cygnus. The handle points to Arcturus, in Bootes.

The stars close to Polaris never set below the horizon at most latitudes in North America, which makes them *circumpolar stars.* They appear to circle around Polaris. Ursa Major is a circumpolar constellation as seen from almost all the Northern Hemisphere. The area of the sky that is circumpolar depends on your latitude. The closer you live to the North Pole, the more of the sky is circumpolar. And in the Southern Hemisphere, the further south your location, the greater the part of the sky that is circumpolar.

Although not circumpolar, Orion is also a distinctive constellation visible in winter, with the three stars that make up its belt pointing to Sirius in Canis Major and Aldebaran in Taurus. Orion also contains the first-magnitude stars Betelgeuse and Rigel, two brilliant beacons in the sky (see Figure 3-2).

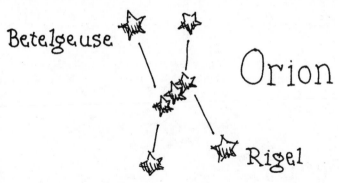

Figure 3-2:
Orion and its
bright stars,
Rigel and
Betelgeuse.

Making friends with the night sky can be done with this book's constellation maps and your own eyes. Just like becoming familiar with the streets of your city helps you find your way faster, knowing the constellations will help you set your sights on the sky objects you want to observe and track their appearance and their movement as you maneuver through a nightly session with the stars.

For a Better View, Use Binoculars or a Telescope

As with any new hobby, it's a good idea to go slow on buying expensive equipment. You should certainly take your time buying a telescope until you've seen several telescopes of different types in action and been able to discuss them with other observers. In the following sections, I offer my advice for choosing the right binoculars or telescope for you.

Binoculars: Best for sweeping the sky

A good pair of binoculars is a must. Buy or borrow binoculars before you get a telescope. They are excellent for many kinds of observation, and if (sigh) you give up astronomy, you can still use them for lots of other purposes.

Binoculars are great for observing variable stars, searching for bright comets and novae, and sweeping the sky just to enjoy the view. You may never discover a comet yourself, but you'll certainly want to spot some of the brighter ones as they appear. Nothing is better for this purpose than a good pair of binoculars.

Figure 3-3 takes you inside a pair of binoculars.

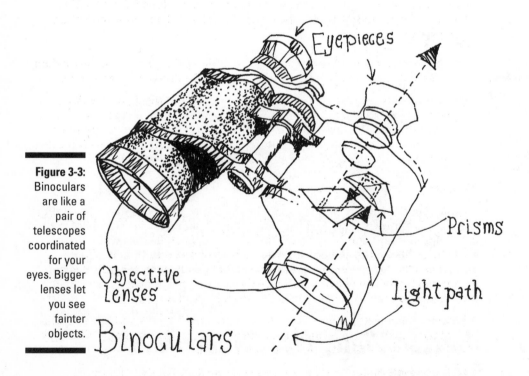

Figure 3-3:
Binoculars are like a pair of telescopes coordinated for your eyes. Bigger lenses let you see fainter objects.

ASTROBABBLE

Good seeing

Turbulence in the atmosphere affects how well you can view a sky object. This causes the stars to twinkle. The term *seeing* is used to describe the conditions of the atmosphere relating to steadiness of the image — *good seeing* is when the air is clear and the image holds steady. You may have better seeing late at night, when the heat of the day has had a chance to dissipate. When seeing is bad, the image tends to "break up" and double stars blur together in telescopic views. Seeing is always worse close to the horizon.

Deciphering the numbers

Binoculars come in many sizes and types. But each pair of binoculars is described by a numerical rating that looks like this: 7 x 35, 7 x 50, 16 x 50, 11 x 80, and so on. (Note that when these are said out loud it is "7 by 35, 7 by 50, 16 by 50, and 11 by 80." Don't say "7 ex 35.") Here's how to decode these ratings:

- ✔ The first number is the optical magnification. A 7 x 35 or 7 x 50 binoculars makes objects look seven times larger than they do to the naked eye.

- ✔ The second number is the *aperture* or diameter of the light-collecting lenses (the big lenses) in the binoculars, measured in millimeters. An inch is about 25.4 millimeters. Thus, 7 x 35 and 7 x 50 binoculars have the same magnifying power, but the 7 x 50 has bigger lenses, which collect more light and can help you see fainter stars than the 7 x 35.

Also, keep these considerations in mind:

- ✔ Bigger binoculars see fainter objects better than small ones, but they are also harder to hold steady and point accurately.

- ✔ Higher magnification binoculars, such as 10 x 50 and 16 x 50, show objects with greater clarity, provided that you can hold them steady enough, but they have smaller fields of view, so finding celestial targets is harder than with lower magnification binoculars.

- ✔ Giant binoculars, 11 x 80, 20 x 80, and on up, are heavy and hard to hold steady; many people can't use them without a tripod or stand. The very biggest, 40 x 150, must be used with a stand.

- ✔ Lots of intermediate sizes are available, such as 8 x 40 or 9 x 56.

Here's my opinion: 7 x 50 is the best size for most astronomical purposes, and certainly the best size to start with. If you purchase binoculars much smaller than 7 x 50, you're really equipping yourself for birdwatching, not astronomy. There are rare exceptions; at least one comet was discovered with only 7 x 35 binoculars. Buy a much larger size than 7 x 50 and you may be investing in a white elephant that you'll rarely use.

You can pay hundreds of dollars or even a few thousand bucks for a good pair of 7 x 50 binoculars, but if you shop around, you can find a perfectly adequate pair for $120, or even less. (A military surplus store is an excellent place to look.) And used binoculars are often a good deal, and much cheaper.

Checking out binocs

Don't buy the binoculars unless you can return them. Here's how to make the basic check to determine whether a pair of binoculars are worth keeping:

- ✔ The image should be sharp across the field of view when you look at a field of stars.

- ✔ You should have no difficulty focusing the binoculars for your eyesight, with a separate adjustment for at least one of the eyepieces (the small lenses that are closest to your eyes when you are looking through the binoculars).

- ✔ When you adjust the focus, it should change smoothly, and the images of stars should be sharp points when in focus and circular in shape when not.

- ✔ Special transparent coatings are deposited on the objective lenses (large lenses) of many binoculars. This feature, called multi-coating, usually results in a clearer, more contrasty view of star fields.

Good binoculars are sold in optical and scientific specialty stores. Some large camera stores have decent binocular selections. But I suggest that you avoid the department stores. You may get low-grade merchandise in some department stores, or pay exorbitant prices for fancy binoculars in others. And you can bet that the salespeople will know less than you do.

Many astronomers buy their binoculars from specialty retailers and manufacturers that advertise in the astronomy magazines. If you must order by mail (or from the Web), find a supplier that is recommended by experienced amateurs whom you meet at an astronomy club or by a staff member at a planetarium.

Reputable makers of binoculars include Bausch & Lomb, Bushnell, Canon, Celestron, Fujinon, Leica, Meade, Nikon, Orion, and Pentax.

Telescopes: When closeness counts

If you are going to look at the craters on the Moon, or the surfaces and cloud decks of the planets, you need a telescope. The same advice goes for observing faint variable stars, or viewing galaxies and the beautiful small glowing clouds called "planetary nebulae," which have nothing to do with planets (see Chapters 11 and 12).

When viewing the Sun, however, or any object crossing in front of the Sun, be sure to read the special instructions in Chapter 10 to protect your eyes and avoid blindness!

Telescopes come in three main classifications:

✔ Refractors use *lenses* to collect and focus light (see Figure 3-4). In most cases, you look straight through a refractor.

✔ Reflectors use *mirrors* to collect and focus light (see Figure 3-5). Reflectors come in different types. In a *Newtonian* reflector, you look through an eyepiece at right angles to the telescope tube. In a *Cassegrain* telescope, you look through an eyepiece at the bottom.

✔ *Schmidt-Cassegrains* and *Maksutov-Cassegrains* use both mirrors and lenses. They tend to be more expensive than comparable reflectors or refractors.

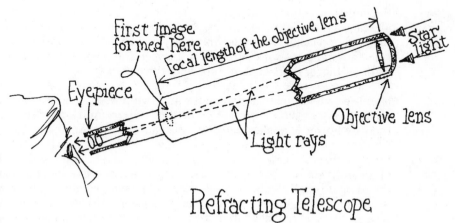

Figure 3-4:
A refracting
telescope.

Eyepiece

Star light

Primary mirror

Prime focus

Reflecting Telescope

Figure 3-5:
English scientist Sir Isaac Newton invented the reflector.

Many varieties are available within these general telescope types. And every telescope used for amateur purposes is equipped with an *eyepiece,* which is a special lens (actually a combination of lenses mounted together as a unit) to magnify the focused image for viewing. When you take photographs, you generally don't use an eyepiece.

Just like a microscope or a camera with interchangeable lenses, almost any telescope can be used with interchangeable eyepieces. Some companies don't make telescopes at all, specializing instead in making eyepieces that can be used on many different telescopes.

Beginners usually buy the highest magnification eyepieces they can, which is a great way to waste your money. I recommend low- and medium-power eyepieces. For a small telescope, observation is usually best with eyepieces rated around 25x or 50x, not 200x or more. (The "x" stands for "times," as in 25 times bigger than what you see with the naked eye.)

If you see a telescope advertised for its "high power," the advertisers may be trying to sell mediocre goods to unwitting buyers. And if a salesperson touts the high power of a telescope, patronize another store.

What limits your view of fine details with a small telescope is not the power of the eyepiece, but the turbulence in the atmosphere or even the shaking of the telescope in the breeze. Therefore, high-power eyepieces can rarely be used. Furthermore, other factors being equal, the higher the power, the smaller the field of view. Often, if you put a high-power eyepiece in a telescope, you'll have difficulty pointing it at and tracking a faint target — or even a bright target.

What color is my universe?

What will you see when you look at a sky object through your binoculars or telescope? Will you see glorious stars, planets, and sky objects in brilliant colors, as shown in the photos in the color section of this book?

Sorry about that, but you're likely to see most stars and sky objects in pale colors. Most stars look white or off-white: yellowish rather than yellow, for example. The colors are most vivid when adjacent stars are in strong contrast, as found in telescopic views of some double stars.

Most photos of sky objects have been color enhanced, traditionally described as having *false color.* False color is not used to gussy-up the universe, which is bright and beautiful all on its own, thank you. Nor is it meant to give a false impression of the deep sky. In fact, the enhancement is done to further the search for truth, much like the stain used on medical slides brings out the detail in the cells and helps identify physical differences and relationships.

Depending on the methods of observation and presentation, photos of the same object can be strikingly different. But they all tell scientists about differences in the structure of the object, what gases it may contain, and what dynamic processes are taking place.

Telescopes are generally mounted in one of two ways:

- ✔ With an *alt-azimuth mount,* you can swivel the telescope up and down and side to side — in azimuth (the horizontal plane) and altitude (the vertical plane). You need to adjust both axes to compensate for the motion of the sky as the Earth rotates. The *Dobsonian* mount is an inexpensive version of the alt-azimuth mount that is used for large amateur reflectors.

- ✔ With the more expensive *equatorial mount,* one axis of the telescope is aligned to point directly at the Celestial North Pole or for Southern Hemisphere viewers, the Celestial South Pole. Once you spot an object, simply turning the telescope around the polar axis will keep it in view. Be sure to polar align the scope for each viewing session.

An alt-azimuth mount is usually steadier, but an equatorial mount is better for tracking the stars as they rise and set.

Remember that the objects you see in the telescope are usually upside down, which is not the case with binoculars. Of course, it doesn't really make much difference in the viewing you're doing, but you should know that top and bottom are reversed when you view through a telescope. Adding a lens to put the image right-side-up would reduce the light coming to the scope and thus reduce the image. Viewed through an equatorially mounted telescope, a star field maintains the same orientation as it rises and sets. But with an alt-azimuth mount telescope, the field rotates during the night, so the stars on top wind up on the side.

Protect your eyes when solar viewing

Taking even the briefest peek at the Sun through a telescope, binoculars, or any other optical instrument is very dangerous unless the device is equipped with a solar filter made by a reputable manufacturer and made specifically for viewing the Sun. And the filter must be properly mounted on the telescope, not jury-rigged.

You must also use a solar filter when viewing planets that are crossing the disk of the sun.

Such planets are said to be *in transit*. Viewing any object against the Sun requires the use of protective viewing techniques. If you have a *Newtonian reflector* or a *refractor*, you can try using projection. See Chapter 10 for specifics about solar viewing and protecting your eyes.

Economical ways to get good scopes

A cheap, mass-manufactured telescope, often called a *drugstore* or *department store* telescope, is usually a waste of money. And it still costs more than a hundred and maybe several hundred dollars.

A good telescope, bought new, may run you the better part of $1,000 and you certainly can pay more. But you can find alternatives:

- ✔ Used telescopes are often for sale, via ads in astronomy magazines or in the newsletters of local astronomy clubs. If you can inspect and test a used telescope and you find what you like, buy it! A well-maintained telescope can last for decades.

- ✔ In many areas, amateurs can observe with the larger telescopes operated by astronomy clubs, planetariums, and public observatories.

The technology of amateur telescopes is advancing at a rapid pace, and yesterday's astronomer's dream can be today's obsolete equipment. Quality and capabilities are going up and price is going down.

Generally speaking, a good refractor gives better views than a good reflector that has the same aperture or telescope size. Aperture or telescope size refers to the diameter of the main lens, mirror, or, in a more complicated telescope, the size of the unobstructed portion of the optics. But a good refractor is much more expensive than a reflector.

A compromise choice

The Maksutov-Cassegrains and Schmidt-Cassegrains are good compromises between the low cost of a reflector and the high performance of a refractor. For many astronomers, they are the preferred telescopes.

As of 1999, the hottest item on the market among small amateur telescopes was the Meade ETX-90/EC, a mightily upgraded version of the older ETX-90, already a highly favored product. Its aperture is 3½ inches, almost the smallest size of any telescope that you should start with. (If you find a good instrument at a good price from a 2-inch aperture on up, especially in a refractor, it's worth considering.)

The basic Meade ETX-90/EC lists at $595, but I recommend the $149 Autostar controller. You'll probably need the $200 field tripod, too. This instrument is so capable that some old hands are denouncing it because it practically runs itself, and will point at any of thousands of sky objects whose RA and Dec are stored in the telescope's little computer (see Chapter 1 if you need more information on RA and Dec). The Autostar can even find moving objects, such as planets, based on stored information.

You definitely don't want to spend this much money until you see the telescope in action at an astronomy club observing meeting or a star party. But the price is no more than you'd pay for a fine camera and an accessory lens or two. You can find larger telescopes for less money; check the ads in current issues of astronomy magazines, but you'll have to invest a lot more effort in learning to use them effectively.

Some name-brand telescopes are sold only through authorized dealers. The dealers tend to have expert knowledge of the telescope business and their advice can be taken with just a wee bit of salt, especially when they carry several competing lines of telescopes.

Key Web sites to browse for telescope product information are:

- ✔ Celestron, for many years the favorite manufacturer for thousands of astronomers, at `www.celestron.com`
- ✔ Meade Instruments Corporation at `www.meade.com`
- ✔ Orion Telescopes & Binoculars at `www.telescope.com`

A Plan for Dipping into Astronomy

I recommend that you get into the astronomy hobby gradually, investing as little money as possible until you're sure about what you want to do. Here's the plan for acquiring both basic skills and needed equipment:

1. **If you have a late-model computer, invest in an inexpensive planetarium program. Start making naked-eye observations at dusk on every clear night, and before dawn if you're an early riser.**

   To plan your observations of planets and constellations, just rely on the weekly sky scenes at the *Sky & Telescope* Web site. If you don't have a suitable computer, plan your observations based on the monthly sky highlights in *Astronomy* or *Sky & Telescope* magazines.

2. **After a month or two of familiarizing yourself with the sky and finding how much you like it, invest in a serviceable pair of 7 x 50 binoculars.**

3. **As you familiarize yourself with the bright stars and constellations, invest in a star atlas that shows many of the dimmer ones, as well as star clusters and nebulae.**

   Norton's Star Atlas (by Arthur P. Norton; 19th edition edited by Ian Ridpath and published by Longman Publishing Group, 1998) has been a leading product for generations. But see what else is available in magazine ads and at planetarium and science museum gift shops. Compare scenes in your star atlas with the constellations that you are starting to know; the atlas will show their RAs and Decs, and you'll start to have a good feel for the coordinate system (see Chapter 1 if you need information about using RAs and Decs).

4. **Join an astronomy club in your own area, if at all possible, and get to know the folks who are experienced with a telescope.**

5. **If all is going well, and you want to continue in astronomy — as I bet you do — invest in a well-made, good quality telescope in the 2.5-inch to 4-inch size range.**

   Study the telescope manufacturer Web sites in this section or send for catalogs advertised in astronomy magazines. Better yet, talk to experienced astronomy club members if you can.

If you find that you're enjoying astronomy as much as I think you will, after a few years consider moving up to a 6-inch or 8-inch telescope. Some on the market are less expensive than that 3.5-inch Meade, for example. They may be much harder to use, but you'll be ready to master them. Equipped with a larger telescope, you can see many more stars and other objects than before.

Chapter 4

Checking Out Visitors: Meteors, Comets, and Man-Made Moons

In This Chapter

▶ Finding the fast facts of meteors, meteoroids, meteorites, and more

▶ Making heads and tails of comets

▶ Spotting artificial satellites

See a moving object in the daytime sky? You probably know whether it's a bird, a plane, or Superman. But in the night sky, can you distinguish the light from a meteoroid from that of an Iridium satellite? And among objects that move slowly but perceptibly across the starry background, can you tell a comet from an asteroid?

This chapter defines and explains many of the objects that sweep across the night sky. (The Sun, Moon, and planets move across the sky, too, but in a more stately procession. I focus on them in later chapters.)

Meteors: Wishing on a Shooting Star

No astronomy term is misused more frequently than the word *meteor*. It's often used incorrectly, even by scientists, when *meteoroid* or *meteorite* is the accurate term. Here are the correct meanings:

- ✔ A *meteoroid* is a small, solid object in space, usually a fragment from an asteroid or comet, orbiting the Sun. Some rare meteoroids are actually rocks blasted off Mars and Earth's Moon.

- ✔ A *meteor* is the flash of light produced when a naturally occurring small, solid object (a meteoroid) enters Earth's atmosphere from space; people often call meteors "shooting stars" or "falling stars."

- ✔ A *meteorite* is a solid object from space that has fallen to the surface of Earth.

Check yourself for space dust

If a *micrometeorite* (a meteorite so small that it must be viewed through a microscope) is found, it may be a particle that began as a cometary meteoroid, or it may be a very small asteroidal meteoroid.

Micrometeorites are so small that they don't create enough friction to burn up or disintegrate in the atmosphere, so they sift slowly down to the ground. Chances are, you have one or two pieces of this space dust in your hair right now, but they would be almost impossible to identify because they would be lost among the millions of other microscopic particles there (no offense).

Scientists collect micrometeorites by flying ultra-clean collecting plates on high-altitude jet aircraft. And they drag magnetized rakes, which pick up micrometeorites made of iron, through the mud in the sea bottom.

If a meteoroid runs into Earth's atmosphere, it may produce a meteor (streak of light) bright enough for you to see. If the meteoroid is big enough to hit the ground rather than disintegrate in midair, it becomes a meteorite. Many people hunt for and collect meteorites, and there's a brisk trade in them.

There are two main kinds of meteoroids, which have different places of origin:

✔ *Cometary meteoroids* are fluffy little dust particles that are shed by comets.

✔ *Asteroidal meteoroids,* which range in size from microscopic particles to boulders, are literally chips from asteroids, the so-called minor planets, which are rocky bodies that orbit the sun (and which I describe in Chapter 7).

When you go to the science museum and see a meteorite on display, you're examining an asteroidal meteoroid that fell to Earth (or, in rare cases, a rock that fell on Earth after being knocked off the Moon or Mars by a larger impacting body). It may be made of stone, iron (actually an almost rustproof mixture of nickel and iron), or both. Showing rare simplicity (for once), scientists call these three types of meteorites, respectively, *stony, iron,* and *stony-iron meteorites.*

Sporadic meteors, fireballs, and bolides

When you are outdoors on a dark night and see a "shooting star" (the flash of light from a random, falling meteoroid) that's a *sporadic* meteor. But if many meteors appear that night, all seeming to come from the same place among the stars, then you are witnessing a *meteor shower*. Meteor showers are among the most enjoyable sights in the heavens, so I devote the next section to them.

A meteor that's dazzlingly bright, is a *fireball*. Although a fireball has no official definition, many astronomers consider a meteor that looks brighter than Venus to be a fireball. However, Venus may not be visible at the time that you see the bright meteor. So how can you decide whether you're seeing a fireball?

Here's my rule for identifying fireballs: If people facing the meteor all say "Ooh" and "Ah" (everyone tends to shout when they see a bright meteor), it's just a fairly bright meteor. But if people who are *facing the wrong way* see a momentary bright glow in the sky or on the ground around them, it's the real thing. To paraphrase an old Dean Martin tune, when the meteor hits your eye like a big pizza pie, that's a fireball!

Fireballs are not very rare. If you watch the sky regularly on dark nights, for a few hours at a time, you'll probably see a fireball about twice a year. It's *daylight fireballs* that are very rare. If the Sun is up and you see a fireball, it's a lucky sighting. That's one tremendously bright fireball. When daytime fireballs happen, they almost always are mistaken for an airplane or missile on fire and about to crash.

Any very bright fireball (approaching the brightness of the half Moon or brighter) or any daylight fireball represents a possibility that the meteoroid producing the light will make it to the ground. Freshly fallen meteorites are often of considerable scientific value, and they may be worth good money, too. If you see a fireball that fits this description, write down all the following information so that your account can help scientists find the meteorite:

1. **Note the time, according to your watch.**

   At the earliest opportunity, check how fast or slow your watch is running.

2. **Record exactly where you are.**

   Chances are good that you don't have a Global Positioning System receiver handy to take a reading of your location, but you can make a little sketch showing where you were standing when you saw the fireball — note roads, buildings, big trees, or any other landmarks.

3. **Make a sketch of the sky, showing the track of the fireball with respect to the horizon as you saw it.**

   Even if you're not sure whether you were facing southeast or north-northwest, your sketch of your location and the fireball track will help scientists determine the trajectory of the fireball, and where the meteoroid may have landed.

After a daylight fireball or a very bright night fireball, interested scientists advertise for eyewitnesses. They collect information like that specified in the preceding list. Then by comparing the accounts of persons who viewed the fireball from different locations, they can home in on the area where it most likely fell to Earth. Even a brilliant fireball may be produced by a small stone, one that would fit easily in the palm of your hand. So scientists need to narrow down the search area in order to have a reasonable chance of finding it.

A *bolide* is a fireball that is seen to explode or that produces a loud noise even if it doesn't break apart. At least, that's how I define it. Some people use "bolide" interchangeably with "fireball." (There is no official agreement on this term, and you can find different definitions in even the most authoritative sources.) The noise that you may hear is the sonic boom from the meteoroid, which is falling through the air faster than the speed of sound.

When a fireball breaks apart, you see two or more bright meteors at once, very close to each other and heading the same way. The meteoroid that produces the fireball has fragmented, probably from aerodynamic forces, just like an airplane falling out of control from high altitude sometimes breaks apart even though its fuel has not exploded.

Often a bright meteor leaves a luminous track behind it. The meteor lasts a few seconds or less, but the shining track — or *meteor train* — may persist for tens of seconds, or even minutes. If it lasts long enough, it becomes distorted by the high-altitude winds, just as the skywriting from an airplane above a beach or stadium is gradually deformed by the wind.

More meteors are seen after midnight than before midnight because, from midnight to noon, you are on the forward side of the Earth, where our planet's plunge through space sweeps up meteoroids. From noon to midnight, you are on the backside, and meteoroids have to catch up with Earth in order to enter the atmosphere and be visible. The meteors are like bugs that splatter on your auto windshield. You get many more on the front windshield as you drive down the highway than on the rear windshield, because the front windshield is driving into bugs, and the rear windshield is driving away from bugs.

Meteor showers — sometimes the best sky sights of the year

Normally, only a few meteors per hour are visible, more after midnight than before, and (for observers in the Northern Hemisphere) more in the fall than in the spring. But on certain occasions every year, you may see 10, 20, or even 50 or more meteors per hour in a dark Moonless sky, far from city lights. These are the occasions of *meteor showers*, when Earth passes through a great ring of billions of meteoroids that runs all the way around the orbit of the comet that shed them. (I discuss comets in detail later in this chapter.) Figure 4-1 illustrates how meteor showers happen.

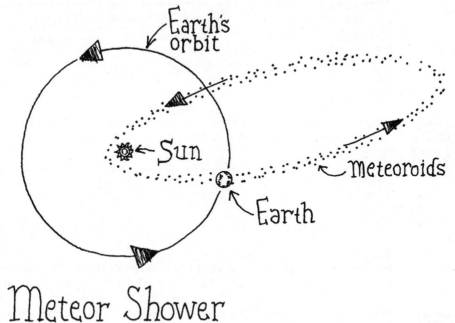

Earth's orbit

Sun

Meteoroids

Earth

Meteor Shower

Figure 4-1: Earth's path crossing a belt of meteroids creates a shower of meteors.

The place where shower meteors seem to come from is called the *radiant*. The most popular meteor shower is the Perseids, which at their peak produce as many as 80 meteors an hour. (The Perseids get their name because they seem to streak across the sky from the direction of the constellation Perseus, the Hero. Meteor showers are usually named for constellations or bright stars (such as Eta Aquarii) that are seen near their radiants.)

A few other meteor showers produce as many or more meteors as the Perseids, but fewer people take the time to observe them. The Perseids come on balmy nights in August, often perfect for skywatching, while the other leading meteor showers — the Geminids and Quadrantids — streak across the sky in the frigid months of December and January, respectively, when the weather is worse and observers' ambitions are limited.

Table 4-1 lists the top annual meteor showers. The dates in the table are the nights when the showers usually reach their peak. Some showers go on for days, and others for weeks, raining down meteors at lower rates than the peak values. But the Quadrantids may last for just one night, or only a few hours.

Table 4-1	Top Annual Meteor Showers	
Shower Name	*Approximate Date*	*Meteor Rate (Per Hour)*
Quadrantids	January 3–4	90
Lyrids	April 21	15
Eta Aquarids	May 4–5	30
Delta Aquarids	July 28–29	25
Perseids	August 12	80
Orionids	October 21	20
Geminids	December 13	100

The Quadrantids' radiant is in the northeast corner of the constellation Bootes, the Herdsman. They are named for a constellation found on 19th-century star charts but that is no longer officially recognized. In addition to having lost the constellation for which they were named, the Quandrantids also seem to have lost the comet that spawned them — their origin is still an astronomical mystery.

The Geminids may be the only meteor shower associated with the orbit of an asteroid rather than a comet. However, the "asteroid" is probably a dead comet, which no longer puffs out gas and dust to form a head and tail. (I discuss comets in the next section.)

The Leonids are an unusual meteor shower that occurs around November 17 every year, usually to no great effect. But every 33 years, many more meteors are present than usual, perhaps for several successive Novembers. Huge numbers of Leonids were seen in November of 1966, and large numbers were predicted for 1999, 2000, and maybe the next year, too.

You will almost never see as many meteors per hour as listed in tables such as Table 4-1. The official meteor rates are defined for exceptional viewing conditions, which few people experience anymore. But meteor showers vary from year to year, just like rainfall. Sometimes, people do see as many Perseids as listed. On rare occasions, they see many more than expected. That's why keeping accurate records of the meteors that you count can be helpful to the scientific record.

To track meteors, you need a watch, a notebook, a pen or pencil to record your observations, and a dim flashlight to see what you're writing.

Enter the red light district

The best light for astronomical observations is a red flashlight, which you can purchase or make from an ordinary flashlight by wrapping red transparent plastic around the bulb. Some astronomers paint the lamp with a thin coat of red nail polish. If you use a white light, you'll dazzle your eyes and be unable to see the fainter stars and meteors for 10 to 30 minutes, depending on the circumstances. Letting your vision adjust to the dark is called getting *dark adapted* and is a step you need to take every time you observe the night sky.

The best way to watch and count meteors is to recline on a lounge chair. (You can do pretty well just lying on a blanket with a pillow but you're more likely to fall asleep in that position and miss the best part of the show.) Tilt your head so that you are looking slightly more than halfway up from the horizon to the zenith (see Figure 4-2). That's the optimum direction for counting meteors.

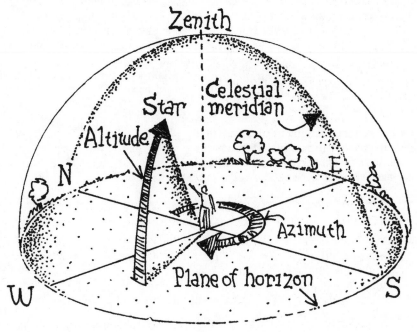

Figure 4-2:
Tilt your head halfway between the horizon and the zenith for optimum meteor viewing.

Photographing meteors and meteor showers

A clear, dark night with no Moon is a good time to try to photograph meteors. For the best results, use an old-fashioned, manually operated 35-mm camera — or a modern camera that can be set for totally manual operation. Then follow these steps:

1. **Use an ordinary lens, not a zoom or telephoto lens, and set it for a distance of infinity.**

2. **Set the f/number of the lens at its smallest value.**

   Use a lens that can be set to f/5.6 or smaller — the smaller the better.

3. **Use a film with a speed of ISO 400.**

   Experts often prefer black-and-white film, but color works fine and is easier and, nowadays, often cheaper to have developed.

4. **Set your camera on a tripod and point the camera about halfway up the sky or a little higher, facing whichever direction has the least interfering sky glow (light pollution) from city or other lights.**

5. **Set the camera for a time exposure and leave the shutter open for 10 to 15 minutes. Then close the shutter, advance the film, and take another exposure.**

If a fireball passes through the part of the sky that the camera is facing, however, make a note of the time at once and close the shutter immediately. Then begin a new exposure.

6. **When you have the film developed, tell the photo processor to "Print all negatives."**

   The operators of photo-processing machines often skip over sky photos, which may appear to be wasted or bad exposures to the nonastronomer.

Photograph a meteor shower the same way that you do a single meteor. But for the best pictures, wait until the radiant (constellation or area from which the meteor shower appears to be coming) is well above the horizon, say 40 degrees or more, before pointing your camera toward it. Then, if you capture several shower meteors on the same exposure, their tracks will resemble the spokes in a bicycle wheel, all pointing back toward the same place, the radiant.

Note on altitudes: The overhead point or zenith is at altitude 90 degrees, and the horizon is altitude 0 degrees, so halfway from horizon to zenith is 45 degrees altitude, two-thirds the way up is 60 degrees, and so on.

You don't have to face toward the radiant when you are observing a meteor shower, although many people do. The meteors will streak all over the sky, and their visible paths may begin and end far from the radiant. But you can visually extrapolate the meteors' paths back in the direction from which they seem to come, and the paths point back to the radiant. That's how you can tell a shower meteor from a sporadic one.

If you do face the radiant, however, you will see some meteors that seem to have very short paths, even though they are fairly bright. The short-looking paths are because they are coming almost right at you. Fortunately, the shower meteoroids are microscopic and won't make it to the ground.

For an excellent beginner's guide to meteor observing, forms to submit your meteor counts, and special forms for reporting fireballs, visit the North American Meteor Network at their Web.InfoAve.Net/~meteorobs Web site. Check out Gary Kronk's meteor and comet site at comets.amsmeteors.org and the International Meteor Organization site at www.imo.net.

Comets: The Lowdown on Dirty Ice Balls

Comets, great blobs of ice and dust that slowly track across the sky looking like fuzzy balls trailing gassy veils, are popular visitors from the depths of the solar system. They never fail to attract interest. Every 75 to 77 years, the best-known, Halley's Comet, returns to the vicinity of Earth and its Sun. If you missed its appearance in 1986, try again in 2061! If you're impatient, you can see other comets in the meantime. Often a less famous comet, such as the recent comet Hale-Bopp, is much brighter than Halley's.

Many people confuse meteors and comets, but you can easily distinguish them: A meteor lasts for seconds; a comet is visible for days, weeks, or even months. Meteors flash across the sky because they are falling overhead, within 100 miles or so of the observer. Comets seem to crawl across the sky because they're seen at distances of many millions of miles. Meteors are common; comets that are easily visible with the naked eye come only once a year or less, on average.

Historically, astronomers described comets as having a head and tail or tails. Later, a bright point of light in the head was named the *nucleus*. Today, we know that the nucleus is the true comet — the so-called dirty ice ball, a stuck-together mixture of water ice, other frozen gases (such as the ices of carbon monoxide and carbon dioxide), and solid particles — the dust or "dirt," shown in Figure 4-3. The other features of a comet are just emanations that stem from the nucleus.

Heads or tails: The structure of a comet

A comet far from the Sun is only the nucleus; it has no head or tail. The ice ball may be dozens of miles in diameter, or just a mile or two. That's pretty small by astronomical standards, and because the nucleus shines only by the reflected light of the Sun, a distant comet is very faint and hard to find.

Images of Halley's nucleus, from a European Space Agency probe that passed very close to it in 1986, showed that the lumpy, spinning ice ball had a dark crust, like the tartufo dessert (balls of vanilla ice cream, coated with chocolate) served in fancy restaurants. Comets are not so tasty, but are real treats to the eye. Here and there on Halley's nucleus, plumes of gas and dust from geyser-like vents or holes sprayed into space where the Sun had barely warmed the surface. Some crust!

As a comet gets closer to the Sun, solar heat vaporizes more and more of the frozen gas, and it spews out into space, blowing some dust out, too. The gas and dust form a hazy, shining cloud around the nucleus. This cloud is the *coma* (a term derived from the Latin for "hair," not the common word for an unconscious state). Almost everyone confuses the coma with the head of the comet, but the head, properly speaking, consists of both the coma and the nucleus.

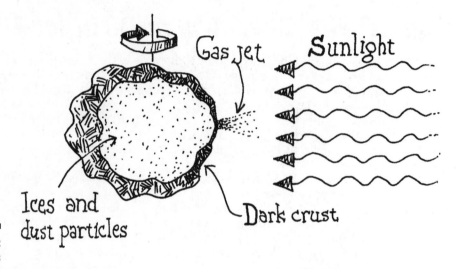

Gas Jet   Sunlight

Ices and dust particles

Dark crust

Dirty Snowball Comet Model

Figure 4-3:
A comet is really just a dirty ice ball.

The glow from a comet's coma is partly the light of the Sun, reflected from millions of tiny dust particles, and partly emissions of faint light from atoms and molecules in the coma.

The dust and gas in a comet's coma are subject to disturbing forces that can give rise to a comet's tail or two.

The pressure of sunlight pushes the dust particles in a direction opposite the Sun (see Figure 4-4), producing the comet's *dust tail.*

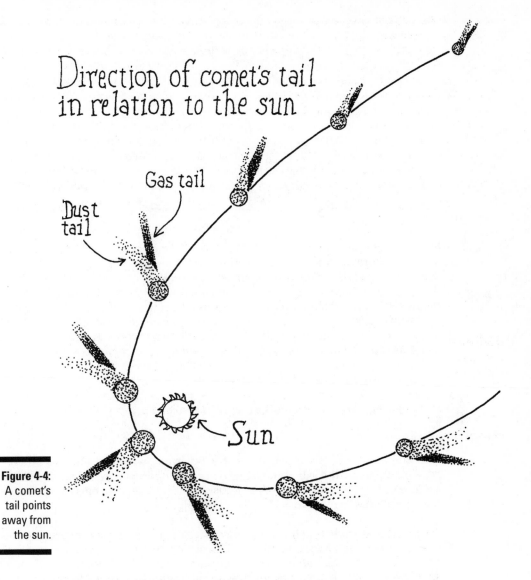

Direction of comet's tail
in relation to the sun

Gas tail

Dust
tail

Sun

Figure 4-4:
A comet's
tail points
away from
the sun.

The dust tail shines by the reflected light of the Sun and has these
characteristics:

✔ A smooth, sometimes gently curved appearance

✔ A pale yellow color

Coma again?

The first rule of viewing comets is: Get out of town! Although a comet's nucleus might be just 5 or 10 miles in diameter, the coma that forms around it can reach tens of thousands, or even hundreds of thousands, of miles across. The gases are expanding, like a puff of smoke from a cigarette. As they thin out, they fade and can't be seen any more. So the size of a comet's coma doesn't just depend on how much stuff is shed by the comet, it also depends on the sensitivity of the human eye or the photographic film (or electronic detector) that's used to observe it. And the apparent size of the coma also depends on the darkness of the sky in which you view it. A bright comet looks a lot smaller downtown than out in the country where skies are dark.

Some of the gas in the coma becomes *ionized,* or electrically charged, when struck by ultraviolet light from the Sun. In that state, the gases are subject to the pressure of the *solar wind,* an invisible stream of electrons and protons that pours outward into space from the Sun (see Chapter 10). The solar wind pushes the electrified cometary gas out in a direction that is also roughly opposite the Sun, forming the comet's *ion* or *plasma tail*. The plasma tail is like a wind sock at an airport: It shows astronomers who view the comet from a distance which way the solar wind is blowing at the comet's point in space.

In contrast to the dust tail, a comet's plasma tail has

- A stringy, sometimes twisted or even broken appearance
- A blue color

Now and then, a length of plasma tail breaks from the comet and flies off in the direction that the tail points. The comet then forms a new plasma tail, like a lizard that grows a new tail when it loses its first one. The tails of a comet can be millions to hundreds of millions of miles long.

When a comet is headed inward toward the Sun, its tail or tails stream behind it. When the comet has rounded the Sun and heads back toward the outer solar system, the tail still points away from the Sun, so the comet now follows its tail! The comet behaves to the Sun as an old-time courtier did to his emperor: Neither turns nor turned his back on his master. The comet in Figure 4-4 could be going clockwise or counterclockwise, but either way, the tail always points away from the Sun.

The coma and tails of a comet are just a vanishing act. The gas and dust shed by the nucleus to make the coma and tails are lost to the comet forever — they just blow away. By the time the comet has traveled far beyond the orbit of Jupiter, where most comets come from, the comet is just a bare nucleus again. But the dust it lost may produce a meteor shower some day, if it crosses Earth's orbit.

"Comets of the century" and all that

Every few years, a comet is sufficiently bright and well placed in the sky that it can be easily seen with the naked eye and small binoculars. I can't tell you when one of those comets is coming, because the only comets whose returns astronomers can accurately predict in the near future are small ones that don't get very bright. Nearly all bright, exciting comets are discovered rather than predicted.

Halley's Comet is the only bright comet whose visits can be accurately predicted, but it doesn't come around very often. Its appearance in 1910 was widely heralded, and everyone got a look. But an even brighter comet, the Great Comet of 1910, also came by that year, and no one had predicted it. So you just have to keep looking up. Monitor the astronomy magazines and the Web sites at the end of this section for reports of new comets and then follow the directions to view them. Or be the first to spot and report a new comet, and it will be named after you.

Every 5 or 10 years, a comet is so bright that it is hailed as "The comet of the century." People have short memories. But stay interested, and you may have your chance:

- ✔ In 1967, Comet Ikeya-Seki was visible in broad daylight next to the Sun, if you held your thumb up to block the bright solar disk. I'll never forget that sight, or my sunburned thumb.

- ✔ In 1976, Comet West was visible to the naked eye in the night sky over downtown Los Angeles, one of the worst places to see celestial objects that I know of.

- ✔ In 1983, Comet IRAS-Iraki-Alcock could be seen (with the naked eye) actually moving in the night sky. (Most comets move so slowly across the stars that you may have to wait an hour or more to notice the effect.)

And in the 1990s, the bright comets Hyakutake and Hale-Bopp appeared out of the blue and were seen by millions of people worldwide.

Web sites galore offer information on currently visible comets and photographs of them from amateur and professional astronomers. Most of the time, the current comets are too dim for any but advanced amateur telescopes. Here are three of the best comet Web sites; check them regularly to make sure that you're not missing anything new:

- ✔ Comet Observation Home Page at NASA's Jet Propulsion Laboratory, `encke.jpl.nasa.gov`

- ✔ The Current Comets page, with histories and observations, at `medicine.wustl.edu/~kronkg/current_comets.html`

- ✔ *Sky & Telescope*'s Comet Page, which offers hints on how to observe and photograph comets, at `www.skypub.com/sights/comets/comets.html`

Your hunt for the Great Comet

Finding a comet isn't difficult, but finding your first one can take years and years. The noted contemporary comet hunter, David Levy, scanned the sky systematically for 9 years before he found his first comet. Since then, he's found over 20 more.

The best telescope to use in comet searching is a "short focus" or "fast" telescope, meaning one whose catalogue specifications include a low f-number (like the f-number of a camera lens) — f/5.6 or better yet, f/4. And you need to use a low-power eyepiece, such as 20x to 30x. The whole idea of the low f/number and the low magnification is to view as large an area of the sky as possible with your telescope. The bright comets that you may be able to discover are few and far between.

You can search for unknown comets in two ways: the easy way and the systematic way.

Finding comets the easy (and haphazard) way

The easy way to search for comets is to make no effort at all. Just be on the lookout for fuzzy patches when you are staring through your binoculars or telescope at stars or other objects in the night sky. Scan the sky for a fuzzy spot (as opposed to stars, which are sharp points of light if your binoculars are in focus). Then check your star atlas to see whether anything at that location is *supposed* to look fuzzy, such as a nebula or a galaxy.

Most important of all, wait a few hours; if the Sun rises or clouds move in the way, blocking your view, look again on the next night. If the object is a comet, it will have moved a bit against the background stars. And, if it's bright enough, the comet may have a tail, which is a dead giveaway.

Finding comets the systematic way

The systematic way to search for comets is based on the following precept: Comets are best found where they are brightest, meaning as close to the Sun as possible, and are best seen where the sky is darkest, which is as far from the Sun as possible. (Even though the Sun has set, the sky in the west remains brighter than the rest of the sky for a considerable time; before dawn, the eastern sky is brighter than the rest of the sky, starting well before sunrise.)

As a compromise between as far from the Sun and as close to the Sun as possible, look for comets in the east before dawn over the part of the sky that's

✔ At least 40 degrees from the Sun (which is below the horizon)

✔ No more than 90 degrees from the Sun

Remember that it's 360 degrees all the way around the horizon, because the horizon is a circle. So 90 degrees is one quarter of the way around the sky.

What's in a name?

If you discover a comet, the comet will be named after you and possibly the next person or two who report it.

If you discover an asteroid, you can recommend someone else that the asteroid will be named for, but not yourself.

If you discover a meteor, there's no time to name it before it's gone. You can try shouting "John," but the name won't catch on and you may attract undue attention. The only meteors that get named are the spectacular ones that are seen by thousands of people in the geographic area where they appear. They get names such as the "Great Daylight Fireball of August 10, 1972."

If you discover a meteorite, it will be named for the town or other local area where you picked it up. The meteorite belongs to the owner of the land where you find it, and if it's U.S. government land, such as a national park or forest, it goes to the Smithsonian Institution.

A desktop planetarium program can help you map out the regions of the constellations that fit this bill for any given night of the year. And, of course, you can look for comets in the west at dusk while following the same two rules. In my experience, the first few "comets" that you discover will be the contrails from jet airplanes, which catch the rays of the Sun at their high altitude, even though the Sun has set at your location.

Start at one corner of the sky area that you plan to check, and slowly sweep the telescope across the area. Then move the telescope slightly up or down and scan the next strip of sky in your search area. You can make every scan from left to right, or you can scan back and forth *boustrophedonically* (a term from classical times that refers to the plowing of a field with oxen; they plow the first furrow in one direction and then come back across the field, plowing in the opposite direction).

It's a lot easier to impress your friends by telling them of your boustrophedonic comet search project than to actually discover a comet. It'll give your ego a boost (unless your friends decide that you are plowed). When you've discovered a comet, follow the directions on the Web site of the International Astronomical Union's Central Bureau for Astronomical Telegrams (which doesn't use telegrams any more) and report it by e-mail. The site is at cfa-www.harvard.edu/iau/cbat.html.

False alarms are not appreciated, so try to get a stargazing friend to check out your discovery before you spread the word. If the find checks out, you — as the amateur discoverer of a comet — may be eligible for a cash share of the Edgar Wilson Award, described on the Central Bureau Web site (cfa-www.harvard.edu/iau/special/EdgarWilson.html).

But even if you never discover a comet, and most astronomers never do, you can enjoy comets that others discover.

Artificial Satellites: Astronomers Love 'em and Hate 'em

An artificial satellite is something that people build and launch into space, where it orbits Earth. These satellites show us the weather, monitor El Niño, relay network television programs, and stand guard against intercontinental missile launches by hostile powers. They can also be used for astronomy.

The Hubble Space Telescope is an artificial satellite, and astronomers love it. It gives us unparalleled views of the stars and distant galaxies, and lets us view the universe in ultraviolet and infrared light that is blocked by the thick layers of Earth's atmosphere beneath Hubble.

But artificial satellites can also catch the rays of not only the setting Sun but the Sun that has already set for observers at ground level, and so they can flash light across the dark sky where an astronomer is making a time exposure photograph of faint stars. This interference is not appreciated. Worse yet, some artificial satellites broadcast at radio frequencies that interfere with the "big dish" radio antennas that astronomers use to receive the natural radio emanations from space. Those radio waves may have traveled for 5 billion years from a quasar, or they may have taken 5,000 years to reach Earth from another solar system in the Milky Way, possibly bearing a greeting from benevolent aliens who would like to send us the cure for cancer. But just when they arrive, there's the blaring tone and strident modulations from a satellite passing over the observatory. We might never know.

So astronomers love satellites when they do something good for us, and hate them when they interfere with our observations.

But to make the best of a bad thing, amateur astronomers have become enthusiastic viewers and photographers of artificial satellites as they pass overhead.

Observing artificial satellites

Hundreds of operating satellites are orbiting Earth, along with thousands more pieces of orbiting space junk — nonfunctional satellites, upper stages from satellite launch rockets, pieces of broken and even exploded satellites, and tiny paint flakes from satellites and rockets.

On the ground, the space shuttle is a manned rocket, but in space, it's an artificial satellite.

You may glimpse the reflected light from any of the larger satellites and space junk, and powerful defense radar tracks even very small pieces.

The best way to get started at observing artificial satellites is to look for the big ones — such as NASA's International Space Station and the space shuttle when it's in orbit in space — and to look for the bright flashing ones, the many Iridium communication satellites.

Looking for a big or bright artificial satellite can be reassuring to the beginning astronomer. Predictions of comets and meteor showers are often mistaken. The comets always seem to be fainter than you expect, and often there are fewer meteors than advertised. But artificial satellite viewing forecasts are usually right on. You can amaze your friends by taking them outside on a clear early evening, glance at your watch, and say "Ho hum, the International Space Station should be coming over about there [point in the right direction as you say this] in just a minute or two." And it will!

✔ A satellite such as the Hubble Space Telescope or the International Space Station generally appears in the evening as a point of light, moving steadily and noticeably from west to east in the western half of the sky. It's much too slow to be mistaken for a meteor, and it moves much too fast to be confused with a comet. You can see it easily with the naked eye, so it's much too bright (and also too fast) to be an asteroid.

Sometimes, you might confuse a high-altitude jet plane with a satellite. But take a look through your binoculars. If you're viewing a plane, you should be able to distinguish running lights, or even the silhouette of the plane against the dim illumination of the night sky. And when your location is quiet, you may be able to hear the plane. You can't hear a satellite.

✔ An Iridium satellite is a wholly different viewing situation: It usually appears as a moving streak of light that gets remarkably bright and then fades out after several seconds. It moves much slower than a meteor. And an Iridium flash is often brighter than Venus, second in brilliance only to the Moon in the night sky. It's caused by the Sun, below your horizon, reflecting off one of the door-sized, flat, aluminum antennas on the satellite. At star parties, people cheer when an Iridium flashes, just like when folks see a fireball. Some Iridium flashes can even be seen in daylight.

And consider this: There are more than 60 Iridium satellites. They interfere with astronomy, and astronomers wish we were rid of them, but at least they provide entertaining flashes that are readily observable.

Finding satellite viewing predictions

Some newspapers and TV weather persons give daily or occasional forecasts for viewing satellites from the local area. However, you can get more detailed information whenever you want it by consulting these sites:

✔ For the International Space Station, *Sky & Telescope* offers observing predictions for 500 cities around the world at www.skypub.com/sights/satellites/satellites.html.

> ✔ For Iridium communication satellites, the most convenient forecasts are available from the German Space Operations Centre (GSOC) at www.gsoc.dlr.de/satvis.

You need to know the latitude and longitude of your observing site to use the Iridium forecasts, but they are easy to determine. The GSOC Web site has a list of these coordinates for 1,500 cities and towns. If you're not on that list or want a more accurate set of coordinates for the block where you live, you can use a handy free service of the U.S. Census Bureau: The U.S. Gazetteer at www.census.gov/cgi-bin/gazetteer.

This service works for locations in the United States. Go to the Web site and follow these steps:

1. **Enter your five-digit zip code or the name of the town where you live (or that's near where you observe) and click on <u>Search</u>.**

   That step brings up a little Census summary of the town, which will even include a latitude and longitude for the town center.

2. **Look for the line that says "Browse Tiger map of area" and click on the word <u>map</u>.**

   You're in business. An area map appears on your screen.

3. **Figure out where your street is on the map, move your cursor to that position, and click on it.**

   This step re-centers the map at your approximate location.

4. **Now click on the option to zoom in.**

   A more detailed map appears.

5. **Put your cursor on what looks like your own location, at greater magnification, and click there to re-center.**

6. **Now scroll down the page and look for the statement of your latitude and longitude.**

   Or keep re-centering and zooming in for more accurate information.

7. **Save the information on latitude and longitude to use when you check your online Iridium forecasts or report fireballs and other astronomical phenomena viewed from the same location.**

After you've seen some bright artificial satellites, you can try photographing them. Follow the directions in the sidebar, "Photographing meteors and meteor showers," earlier in the chapter. All you need is a camera suitable for taking time exposures, a steady tripod, and some fast film.

Congratulate yourself; you know the night visitors, and you can look forward to their entertainment.

Part II
Once Around the Solar System

The 5th Wave   By Rich Tennant

"I think what you mean, dear, is a 'magnetic' storm."

In this part . . .

Guess what? Men aren't from Mars, and women aren't from Venus. In fact, neither planet can support life as we know it. Venus is too hot, Mars is too cold, and neither has any liquid water. This part explains what the planets of Earth's solar system are really like. Did Mars ever support life? What about Jupiter's moon, Europa? I tell you what is known as of now.

And if you've watched any of those "Oh, no, a big asteroid is headed for Earth!" movies, you may be wondering whether you should be worried. I include a chapter that explains all about asteroids and tells you the truth about the risk of impact.

Chapter 5

Earth and Its Moon

· ·

In This Chapter

▶ Seeing Earth as a planet

▶ Understanding time and seasons

▶ Phasing in on the Moon and observing eclipses

▶ Digging the Moon's craters

· ·

*P*eople often think of planets as objects in the sky, like Jupiter and Mars. The ancient Greeks — and people for centuries thereafter — distinguished between Earth, which they regarded as the center of the universe, and the planets. The planets were little lights in the sky that revolved around Earth.

Today we know better. Earth isn't the center of the universe. It isn't even the center of the solar system. The Sun is. The Moon goes around Earth, along with hundreds of artificial satellites, and that's about it. And joining Earth in the track around the Sun are eight other planets in the solar system, a number of other moons, and a belt of asteroids and other space debris. Nevertheless, as far as we know now, life exists in the solar system only on Earth.

Earth has fallen from its exalted place in human thought as the center of the universe to its true but still significant status: our home planet. And there's really no other place in the solar system that's quite like home.

Earth is what astronomers call a *terrestrial* planet. That's kind of a circular definition, because "terrestrial" means Earthlike. But the scientific meaning is a planet made of rock, orbiting the Sun. There are four terrestrial planets, the ones closest to the Sun. In order of distance from the Sun, they are Mercury, Venus, Earth, and Mars.

Some people consider Earth's Moon a terrestrial planet, and regard the Earth-Moon system as a double planet. For aliens seeking to visit us, that's probably a good idea: "Just head for that yellow-white star in Sector 49,832 of the Orion Arm, in the Milky Way, and home in on the third rock from that Sun; it's a double planet and easy to spot."

Earth: What's So Special about It?

What's special about Earth? Earth is the only planet we know of with

- ✔ **Liquid water at the surface.** Earth has lakes, rivers, and oceans, unlike any other planet. The oceans cover 70 percent of the surface of Earth.

- ✔ **Lots of oxygen in the air.** The air on Earth contains 21 percent oxygen; no other planet has more than a trace of oxygen in its present atmosphere.

- ✔ **Plate tectonics, also known as continental drift.** Earth's crust is composed of huge moving plates of rock; where plates collide, earthquakes occur and new mountains rise. New crust emerges at the mid-ocean ridges, deep beneath the sea, causing seafloor spreading.

- ✔ **Active volcanoes.** Hot molten rock, welling up from deep beneath the surface, forms huge volcanic landforms such as the Hawaiian Islands. Volcanoes erupt somewhere on Earth every day.

- ✔ **Life, intelligent or otherwise.** You be the judge on intelligence, but from one-celled amoebas to bacteria and viruses, to flowers and trees, fish and fowl, insects and mammals, Earth has life in abundance.

There are tantalizing indications that Mars and Venus may once have shared some of these traits with Earth (see Chapter 6). But as far as we know, they don't have these properties now.

Scientists believe that the presence of liquid water on the surface of Earth is one of the main reasons why life is here. You can easily imagine advanced life-forms on other worlds. You can see them on TV and in the movies. But they are all imaginary. Despite some recent claims, scientists don't have convincing evidence for any life, past or present, anywhere but on Earth.

Spheres of Influence: Dividing Earth

Figure 5-1 shows four views of Earth as seen from space. Earth's patterns of land and sea and clouds are clearly visible.

Scientists classify the regions of Earth into

- ✔ The *lithosphere,* which means all of Earth's rock

- ✔ The *hydrosphere,* which is the water in the oceans, lakes, and elsewhere on Earth

- ✔ The *cryosphere,* which is the frozen regions — notably the Antarctic and Greenland ice caps

✔ The *atmosphere,* which is the air from ground level on up for hundreds of miles

✔ The *biosphere,* which consists of all the living things on Earth, in the air and water and underground

So you are part of the biosphere that lives on the lithosphere, drinks from the hydrosphere, possibly skates on the cryosphere or uses it to cool your drinks, and breathes the atmosphere. I don't know anywhere else in space where you can do all that.

In addition to those previously described, one more important subdivision is the *magnetosphere,* which plays an important part in protecting Earth from the negative effects of the Sun, which I discuss in Chapter 10. The magnetosphere is produced by the geomagnetic field, which I explain in the section that follows.

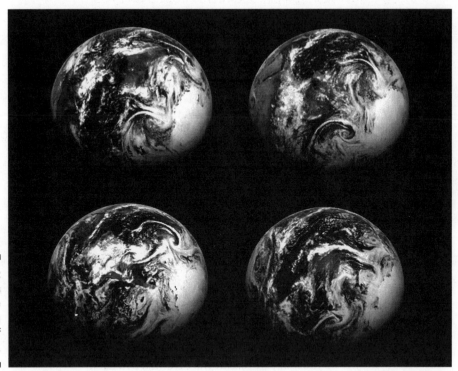

Figure 5-1:
Four views
showing the
changing
face of
Earth.

Photo courtesy of NASA

The magnetosphere: A main attraction

Sometimes called Earth's radiation belts (or the Van Allen radiation belts, named for James Van Allen, the U.S. physicist who discovered them with America's first artificial satellite, Explorer 1), the magnetosphere consists of electrically charged particles, mostly electrons and protons, which bounce back and forth above Earth, trapped in Earth's magnetic field.

Occasionally, some of the electrons escape and rain down on Earth's atmosphere below, striking atoms and molecules and making them glow. That glow is the aurora, known as the *aurora borealis* (northern lights) when viewed in the Northern Hemisphere and the *aurora australis* when viewed in the Southern Hemisphere. See the sidebar at the end of the chapter for more about viewing auras.

Earth's magnetic field

The solid surface of Earth — the part you stand on — is the crust. Beneath the crust are the mantle and then the core. The core is largely iron and nickel and very hot, reaching about 12,000°F (7,000°K) at the center. And the core is layered, too: The outer core is in a molten state, and the inner core is solid.

The extremely high pressure of the overlying layers makes the hot iron in the inner core solidify. As Earth cools down over millions of years in the future, the solid part at the center will increase in size at the expense of the surrounding molten core, like an ice cube growing as the surrounding liquid cools.

Earth's core is far below our ability to dig, but it produces an effect that anyone can observe at the surface. Moving streams of molten iron in the outer core generate a magnetic field that reaches out through the whole planet and far into space. It's called the *geomagnetic field*.

The geomagnetic field

- Makes the compass needle point
- Provides an invisible guidance system for homing pigeons, some migratory birds, and even some ocean-dwelling bacteria
- Forms the magnetosphere far above Earth
- Shields Earth from incoming electrically charged particles from space, such as the solar wind and many cosmic rays

The geomagnetic field is a *global planetary magnetic field*. That means it extends above all parts of Earth and is continuously being generated. Mars, Venus, and the Moon all lack a global magnetic field like Earth's, and this key difference gives scientists information about the cores of those objects. For more on the lunar core, see the "Giant Impact: A theory of the origin of the Moon" section, later in this chapter.

Spreading rock on the seafloor

According to geophysical surveys, patterns of magnetized rock exist in the seafloor on either side of mid-ocean ridges. The rock became magnetized as it cooled from the molten state, trapping and "freezing in" some of Earth's magnetic field that pervaded it as the rock solidified. So the seafloor rock resembles a magnet, with a magnetic field that has strength and direction. After the rock solidified, its magnetic field could change no longer; it's now a fossil magnetic field like a fossil dinosaur that remains forever in the shape it had at death.

The patterns discovered near the mid-ocean ridges consist of stripes of magnetized rock, hundreds of miles long, that parallel the ridges and alternate in direction. One stripe has a north magnetic polarity, like the end of a bar magnet that attracts a north-seeking compass needle, the next stripe has the opposite polarity, and so on.

The alternating stripes of oppositely magnetized rock are due to the new rock emerging from the mid-ocean ridges, cooling and magnetizing, and moving away from the ridges as even newer rock pushes it along. The oppositely magnetized stripes show that the geomagnetic field itself periodically reverses direction, like a bar magnet that is turned 180 degrees at intervals — except that the intervals for the geomagnetic field are probably hundreds of thousands of years.

An unknown process causes the geomagnetic field, generated deep in Earth's core, to reverse every so often. That effect is preserved in the fossil magnetic fields of the rock at the seafloor and in rock on the continents that previously lay beneath the sea.

Why mention all this seafloor stuff in a book on astronomy? Because this unusual property of Earth may correspond to a phenomenon discovered on Mars. As scientists consider the evidence gathered on the various terrestrial planets, including Earth, we find similarities and differences that help us understand them better. That research is called *comparative planetology,* which I cover in more detail in the descriptions of Mars and Venus in Chapter 6.

Time and the Motions of Earth

Nowadays, atomic clocks are used to measure time with great precision. But originally and until modern times, Earth's system of time was based on the turning of Earth.

Earth rotates once on its axis in 24 hours. It turns from west to east (or counterclockwise as visible from above the North Pole). And it orbits the Sun counterclockwise, as visible from space way above the North Pole. The length of the day, 24 hours, is the average time it takes for the Sun to rise and set and rise again. This is called *mean solar time,* and is equivalent to the standard time on your watch.

The length of the day, then, is 24 hours of mean solar time. And there are approximately 365 days in the year, the time that it takes Earth to make one complete orbit around the Sun.

Orbiting for all time

Because Earth is moving around the Sun, the time when you see the Sun rise depends on both the rotation of Earth and Earth's orbital motion.

The Earth turns once in 23 hours 56 minutes and 4 seconds with respect to the stars. That amount of time is called the *sidereal day.* (Sidereal means pertaining to the stars.) Notice that the difference between 24 hours and 23 hours, 56 minutes, and 4 seconds, which is 3 minutes and 56 seconds, is just about $\frac{1}{365}$ of a day. That's no coincidence. It happens because during a day, Earth moves through $\frac{1}{365}$ of its orbit around the Sun.

Astronomers used to depend on special clocks called sidereal clocks, which measured sidereal time by registering 24 sidereal hours during an interval of 23 hours, 56 minutes, and 4 seconds of mean solar time. The sidereal hours, minutes, and seconds are all slightly shorter than the corresponding units of solar time. Using sidereal clocks enabled astronomers to keep track of the stars and point telescopes correctly. But astronomers and you don't need to do that anymore. Computer programs that point telescopes, or that picture the sky on a desktop planetarium, as I describe in Chapter 2, do all of the mathematics for you, so you can simply use the standard time at your location to figure out where different stars and constellations appear in the sky.

On the other hand, it's customary to report astronomical observations in a common system used by astronomers everywhere, which is called *Universal Time (UT)* or *Greenwich Mean Time.* UT is simply the standard time at Greenwich, England. By international agreement, the day begins at Greenwich. (Actually, it begins at midnight in Greenwich, which is zero hours UT.) If you live in the United States, the standard time at your location is always earlier than the time at Greenwich. For example, in New York City, the Sun rises about 5 hours after it rises at Greenwich. When it's 6 a.m. at Greenwich, it's still only 1 a.m. in New York.

A more precisely defined time, Coordinated Universal Time or UTC, which is identical to UT for all practical purposes, is the official international standard.

In the United States, the U.S. Naval Observatory in Washington, DC, is in charge of time. You can get the UTC any time you want it from the U.S. Naval Observatory Web page at `tycho.usno.navy.mil/what.html`, where there's also a place that you can click to compute the Local Apparent Sidereal Time at your observing site. The *Local Apparent Sidereal Time* equals the right ascension (see Chapter 1) of the stars that are on your meridian, the north-south line above you in the sky. A star is highest in the sky and best placed for observation when it is on the meridian.

To convert from UTC to Standard or Daylight Time at any place in the United States, including Hawaii and Alaska or in Samoa, follow the simple instructions at `aa.usno.navy.mil/AA/faq/docs/us_tzones.html`. And to determine the Standard Time Zone that applies at just about any other place in the world, and convert it to Universal Time, consult the World Time Zone Map of Her Majesty's Nautical Almanac Office, at the U.S. Naval Observatory Web site, `aa.usno.navy.mil/AA/faq/docs/world_tzones.html`.

Generally speaking, Daylight Saving Time (called Summer Time in the U.K.) is an hour later than Standard Time at the same geographic location. But not all places observe Daylight Saving Time.

Tilting at the seasons

Teaching students about the cause of the seasons on Earth is about the most frustrating task of any astronomy professor. No matter how carefully the professor explains that the seasons have nothing to do with how far Earth is from the Sun, many if not most students don't buy it. Surveys taken, even at Harvard University graduation, show that bright college graduates think that summer is when Earth is closest to the Sun, and winter is when Earth is farthest from the Sun.

What students forget is that when it's summer in the Northern Hemisphere, it's winter down south. And when it's summer in Australia, it's winter in the United States. But Australia and the U.S. are on the same Earth. Earth can't be farthest from the Sun and closest to the Sun at the same time. It's a planet, not a magician.

The actual cause of the seasons is the tilt of Earth's axis (see Figure 5-2). The *axis*, the line through the north and south poles, is not perpendicular to the plane of Earth's orbit around the Sun. Actually, the axis is tilted by 23½ degrees from the perpendicular to the orbital plane. The axis points north to a place among the stars, in fact near the North Star. (At least in the short term; the axis slowly changes its pointing direction and the North Star in one era will no longer be the North Star in the far future.)

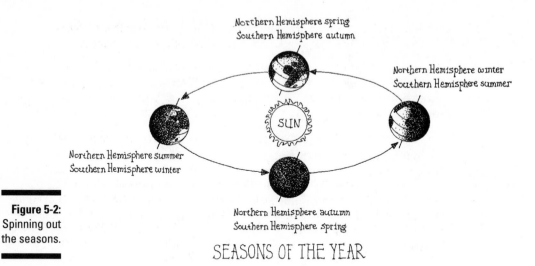

Northern Hemisphere spring
Southern Hemisphere autumn

Northern Hemisphere winter
Southern Hemisphere summer

SUN

Northern Hemisphere summer
Southern Hemisphere winter

Northern Hemisphere autumn
Southern Hemisphere spring

Figure 5-2:
Spinning out
the seasons.

SEASONS OF THE YEAR

At present, the North Star, also called Polaris, is the star Alpha Ursae Minoris, located in the Little Dipper asterism of the Lesser Bear constellation, Ursa Minor. If you are lost at night and want to bear north, set your sights on the Little Dipper (see Chapter 3 for more about finding Polaris).

The axis of Earth points "up" through the North Pole and "down" through the South Pole. When Earth is on one side of its orbit, the axis pointing "up" is also pointing roughly toward the Sun, so that the Sun is high in the sky at noon in the Northern Hemisphere. Six months later, the axis is pointing "up" and pointing roughly away from the Sun. Actually, the axis always points in the same direction in space, but Earth is now on the opposite side of the Sun.

Summer occurs in the Northern Hemisphere when the axis pointing up through the North Pole is pointing roughly at the Sun. When that happens, the Sun at noon is higher in the sky than at other seasons of the year, so it shines more directly on the Northern Hemisphere and provides more heat. At the same time, the axis pointing down through the South Pole is pointing away from the Sun, so the Sun is lower in the sky at noon than at any other season of the year, creating less direct sunlight, so winter occurs in Australia.

There are more hours of sunlight in the summer because the Sun is higher in the sky. It takes longer to rise to that height, and it takes longer to set from that position. The longer hours of sunshine help make Earth warmer at that season of the year.

ASTROBABBLE

As Earth orbits the Sun, the Sun seems to move through the sky, following a circle in the sky called the *ecliptic,* which I mention in Chapter 3. The ecliptic is tilted with respect to the equator by exactly the same angle as Earth's tilt on its axis: 23½ degrees.

- ✔ When the Sun crosses from "below" (south) the equator to "above" (north), it's the first day of spring, called the *vernal equinox.*

- ✔ When the Sun reaches the farthest point north on the ecliptic, it's the *summer solstice.*

- ✔ When it crosses the equator going back down south, it's the start of fall, the *autumnal equinox.*

- ✔ And when the Sun gets as far south as possible on the ecliptic, it's the *winter solstice.*

In the Northern Hemisphere, the summer solstice is the day with the most hours of sunlight during the year, because that's when the Sun gets highest in the sky — taking the longest time to reach that height and come back down to the horizon again. By the same token, in the Northern Hemisphere, the winter solstice is the day with the shortest amount of daylight during the year.

That's the long and the short of time and seasons.

Estimating Earth's age: It's aeons and aeons!

Measuring radioactivity is the only accurate way we have to date very old things on Earth, or in the solar system. Some elements, such as uranium, have unstable forms called *radioactive isotopes.* A radioactive isotope turns into another isotope of the same element, or into a different element, at a rate determined by the half-life of the radioactive substance. If the *half-life* is one million years, for example, half of the radioactive isotope that was originally present will have turned into another substance (called the *daughter isotope*) in one million years, leaving half still radioactive. And half of the remaining half will turn into daughter isotope atoms in another million years. So after two million years, only 25 percent of the original radioactive isotope atoms will still exist. After three million years, only 12½ percent will be left. And so on.

When the original radioactive isotope atoms, called the *parent atoms,* and the daughter atoms are trapped together in a piece of rock or metal, such as a meteorite, scientists can count the atoms' respective numbers to determine how old the rock is. That process is called *radioactive dating.*

By the method of radioactive dating, the oldest rocks on Earth are about 3.8 billion years old. However, Earth is undoubtedly much older than that. Erosion, mountain building, and volcanism (the eruption of molten rock from within Earth, including the formation of new volcanoes) are constantly destroying the rocks at the surface of Earth, so the original surface rocks of Earth are long gone.

Meteorites, however, yield radioactive dates as old as 4.6 billion years. Meteorites are considered debris from asteroids, and asteroids are thought to be debris from the very early solar system, when the planets first formed (see Chapter 7 for more about asteroids).

So scientists think that Earth and other planets are about 4.6 billion years old. Earth's Moon, however, is younger. And that's the next story.

Earth's Moon

The Moon is 2,160 miles (3,476 kilometers) in diameter, slightly more than ¼ the diameter of Earth. Its mass is only ⅛₁ the mass of Earth, and its density is about 3.3 times the density of water, noticeably less than the density of Earth, which is 5.5 times the density of water. The Moon has no meaningful atmosphere, just a trace of hydrogen, helium, neon, and argon atoms — and others in even lesser quantities. It appears to be made of solid rock (see Figure 5-3).

Density is a measure of the amount of mass that is packed into a given volume. If you have two cannon balls of the same size and shape, they have the same volume. But if one ball is made of lead and one is made of wood, the lead ball is heavier. It has a higher density.

Figure 5-3:
The Moon is made of rocks and rilles, craters and dried lava seas — not a speck of cheese in sight.

Photo courtesy of NASA

Although the Moon is less dense than the average density of Earth, it has about the same density as the mantle of Earth, the layer beneath the crust and above the core. The average density of Earth is larger than the density of the mantle because the core is nearly all iron and nickel, which are much denser than rock. Earth's core is denser than the average of the whole Earth, and the mantle is less dense than the average. These differences are very informative, as I explain in the section "Giant impact: A theory of the origin of the Moon."

Phases of the Moon

Except during a lunar eclipse, half the Moon is always in sunlight, and half of it's always in night. But these are not, as people sometimes think, the lunar near side and the lunar far side. Those are the hemispheres that point toward and away from Earth, which are always the same. The lunar halves that are in sunlight and in night are the hemispheres that face toward and away from the Sun. And they are always changing as the Moon moves around Earth (see Figure 5-4).

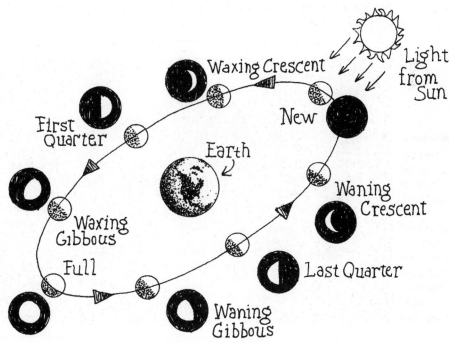

Figure 5-4: The phases of the Moon. Avoid the full Moon when Moonwatching.

Moon's Orbit and Phases of the Moon

New Moon is the beginning of the monthly lunar cycle or *lunation.* At this time, the near side faces away from the Sun. So the near side is the dark side. A few hours or days later, the Moon is a new crescent, or *waxing crescent,* meaning a crescent Moon whose bright area is getting larger. This phase happens as the Moon moves away from the Sun-Earth line while orbiting Earth. Fully half of the Moon is always lit up, facing the Sun, but when it is a crescent Moon, most of this illuminated area faces away from Earth and cannot be seen by us.

As the Moon moves around its orbit, it reaches a point where the Earth-Moon line is at right angles to the Earth-Sun line. At this point, we see a *half Moon,* which astronomers call a *quarter Moon.*

How can a half equal a quarter? It's not possible if you're making change, but it's easy for the astronomer. Half of the lunar near side, the part facing Earth, is lit up, so people call it a half Moon. But the illuminated portion of the Moon that we see is only half the bright hemisphere that is facing the Sun and half of a half is a quarter. Bet your friends that a quarter can be a half. You'll win, and you can pocket the change!

When the illuminated part of the Moon that we can see is larger than the quarter (half) Moon and smaller than the full Moon, it's called a *gibbous Moon.*

People often ask why an eclipse of the Sun doesn't occur every month at new Moon. The reason is that Earth, Moon, and Sun are not usually all exactly on a line at new Moon. When they *are* all on a line, an eclipse of the Sun results. When Earth, Moon, and Sun are all on a line at full Moon, there's an eclipse of the Moon.

When the Moon is on the far side of its orbit, opposite the Sun in the sky, the lunar hemisphere that faces Earth is fully lit. That's *full Moon.* As the Moon continues around its orbit, the illuminated portion gets smaller and the Moon is gibbous again, less than full and more than a quarter Moon. Soon it is a quarter Moon again, called *last quarter.* Next, as it nears the line between the Earth and the Sun, it's a *waning crescent Moon. Waning* means that the crescent is getting smaller. Then it's new Moon, and the cycle of phases starts over again.

The Earth has phases, too, just like the Moon! To see them, head into space and look back at Earth from the Moon. When folks on Earth see a beautiful full Moon, you'll enjoy a "new Earth" and when Earthlings experience new Moon, you'll be lit by a full Earth.

Eclipses of the Moon

Total eclipses of the Moon are not as common as total eclipses of the Sun, but we see them more often. That's because a total eclipse of the Sun is visible only along a narrow band on Earth, the path of totality. But when Earth's shadow falls on the Moon, the eclipsed Moon can be seen from all over half of Earth, wherever it is night.

A lunar eclipse occurs when the full Moon is exactly on the line from the Sun to the Earth. Then the Moon is in Earth's shadow, the *umbra*. It's perfectly safe to look at a lunar eclipse, as long as you don't bump into something in the dark or stand in the road.

During a total eclipse of the Moon, you can still see the Moon, although it is in Earth's shadow (see Figure 5-5). No direct sunlight is falling on the Moon, but some rays of sunlight get bent through Earth's atmosphere around the edges of Earth (as visible from the Moon), and fall on the Moon. That sunlight is strongly filtered as a result of passing through our atmosphere; usually only the red and orange parts of sunlight get through. This effect differs from one lunar eclipse to the next, depending on meteorological conditions and clouds in Earth's atmosphere during each eclipse. The totally eclipsed Moon, therefore, can look a dull orange, an even duller red, or a very dark red. Sometimes, you can barely make out the eclipsed Moon at all.

The upcoming total lunar eclipses for the next 12 years are:

> January 21, 2000
>
> July 16, 2000
>
> January 9, 2001
>
> May 16, 2003
>
> November 8, 2003
>
> May 4, 2004
>
> October 28, 2004
>
> March 3, 2007
>
> August 28, 2007
>
> February 21, 2008
>
> December 21, 2010
>
> June 15, 2011
>
> December 10, 2011

In advance of each of these eclipses, you can find lots of information on the exact times and on the part of Earth where the eclipse will be visible. Look for those data in *Astronomy* and *Sky & Telescope* magazines and on their Web sites (www2.astronomy.com/astro and www.skypub.com), as well as in your local newspaper and television weathercasts.

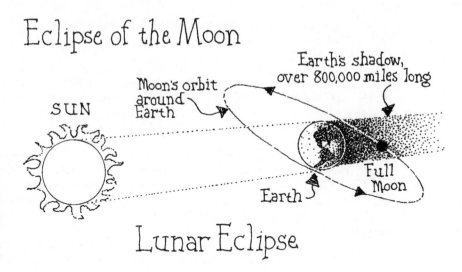

Figure 5-5:
A total
eclipse of
the Moon.

Partial eclipses aren't quite as interesting. During a *partial* eclipse, only some of the full Moon is in Earth's shadow. The Moon just appears to be at a different phase. People who don't know that an eclipse is in progress — nor that it's the time of the full Moon — won't be aware that something special is underway. They will just think that it's a quarter or crescent Moon. But if they keep looking for an hour or so, they'll see the full Moon come out from Earth's shadow. And then they'll know.

Lunar geology

The entire Moon is pockmarked with craters of every size, from microscopic pits to basins hundreds of miles in diameter. The largest is the South Pole-Aitken Basin, which is about 1,600 miles (2,600 kilometers) across. These craters were caused by objects (asteroids, meteoroids, and comets) that struck the Moon, for the most part, very long ago. The microscopic craters, which have been found on rocks brought back from the surface of the Moon, are caused by micrometeorites, tiny rock particles flying through space. All of these craters and basins are known collectively as *impact craters,* to distinguish them from volcanic craters.

There has been volcanism on the Moon, but it took another form than Earth's. There are no *volcanoes,* large volcanic mountains with craters at the top. But there are small volcanic domes, round-topped hills that occur in some volcanic regions on Earth. In addition, sinuous channels on the lunar surface (called *rilles*) appear to be lava tubes, also a common landform in volcanic areas on Earth, such as Lava Beds National Monument in northern California. Most notably, the Moon has huge lava plains that fill the bottoms of the large impact basins. These lava plains are called *maria*, the Latin word for seas.

When you look up and see the Man in the Moon, the dark areas that make up some of his features are the maria.

Some early scientists thought that the maria might be oceans, but they are just dead, dried lava beds. If they were oceans, you would see bright reflections of the Sun from them, just as you do when you look down at the sea from an airplane during the day. The larger bright areas in the Man in the Moon are the *lunar highlands*, which are very heavily cratered areas. The maria have craters, too, but fewer craters per square mile than the highlands, which means that the maria are younger. Huge impacts created the basins where the maria are located. These impacts obliterated preexisting craters. Later, the basins filled with lava from below, wiping out any new craters that had formed after the huge impacts. All the craters seen in the maria now are from impacts that happened after the lava froze.

In the late 1990s, a NASA spacecraft called the Lunar Prospector obtained indirect evidence indicating that there may be frozen water in the bottoms of a few craters near the north and south poles of the Moon, where the Sun never shines. The Sun, at best, is low on the horizon near the poles of the Moon and is blocked by the crater rims from shining on the crater bottoms. This ice may come from comets that struck the Moon long ago, because comets are largely ice and they do hit the Moon and planets now and then. But there is no other water on the Moon.

Sights on the Moon and lunar charts

The Moon is one of the most rewarding objects to observe. It can be seen when the sky is hazy or partly cloudy, and is even visible during the day sometimes. You can see craters with even the smallest telescopes. With a good quality small telescope, you can enjoy hundreds, maybe thousands, of lunar features, including

- **Impact craters:** Round structures caused by the impact of meteoroids and other larger bodies slamming into the Moon; the largest craters are called basins.
- **Maria:** Lava beds that fill the bottoms of the basins.
- **Lunar highlands:** Heavily cratered lunar areas.
- **Rays:** Bright lines that extend radially outward from young, bright impact craters, such as Tycho and Copernicus (see Figure 5-6). They are formed by powdery debris thrown out from the impacts.
- **Rilles:** Winding channels; the lunar name for what probably are lava tubes. The most famous is the Hadley Rille, which was visited by Apollo astronauts.
- **Central peaks:** Mountains of rubble thrown up in the rebound of the lunar surface from the effects of a powerful impact. Central peaks are found in some but not all impact craters.

✔ **Lunar mountains:** The rims of large craters or impact basins, which may have been partly destroyed by subsequent impacts, leaving parts of their walls standing alone as if they were a range of mountains. They are not the type of mountain you see on Earth.

If you want to know which crater, rille, or lunar mountain range is which as you look through your telescope, you need a Moon map or a set of lunar charts. These inexpensive items are available from astronomy and other scientific hobby supply houses, and sometimes from map stores. Some good sources and the Moon maps they offer are

✔ Edmund Scientific (www.edsci.com) sells a full color Moon map poster with feature identifications. The more expensive version (about $10) is better for use with your telescope, because it's printed on plastic. In the cool night air, paper tends to get moist with dew.

✔ Orion Telescopes & Binoculars (www.telescope.com/default.asp) offers a more detailed Atlas of the Moon (about $35), with 76 charts.

✔ The nonprofit Astronomical Society of the Pacific (www.aspsky.org) lists the Rand McNally map of the Moon in its catalog. This map is more than 3 feet on a side and has an index of 1,800 lunar features. It sells for about $9, and you can probably find it in map stores and planetariums, too.

These maps and charts for use at the telescope show only one side of the Moon, the lunar near side.

Crater Copernicus • The Moon
Hubble Space Telescope • WFPC2

PRC99-14 • STScI OPO • J. Caldwell (York University), A. Storrs (STScI) and NASA

Figure 5-6:
Crater
Copernicus
and its rays.

Courtesy of NASA

A view of the far side

Only the lunar near side is visible from Earth. Our view is limited because the Moon is in *synchronous rotation,* meaning that it makes exactly one turn on its axis as it makes one orbit around Earth (the orbital period of the Moon, which is the same as its "day," is about 27 days, 7 hours, and 43 minutes).

Astronomy supply houses and science stores also sell Moon globes, however, that depict the features of the entire Moon, meaning the lunar near side and the *far side*. The lunar far side is not by Gary Larson, but was brought to us by the Russians. The first photographs of the lunar far side were made by the Soviet space program, which snapped pictures using a robotic spacecraft very early in the Space Age. Since then, the Moon has been thoroughly mapped with many different U.S. spacecraft, including the Lunar Orbiters and Clementine. You don't need a chart of the far side to help you observe the Moon, because you can't see the far side.

Make friends with the terminator

For almost anything you want to see on the Moon, the best viewing time is when the object is near the *terminator,* the dividing line between bright and dark. You can see detail best when a lunar formation is just to the bright side of the terminator.

About the worst time to look at nearly anything on the Moon is at full Moon. During full Moon, the Sun is high in the sky on most of the lunar near side, so shadows are few and short. The presence of shadows cast by features on the Moon helps you understand the surface relief — the way landforms extend above or below their surroundings.

During a month, which is approximately the period of time from one full Moon to the next one, the terminator moves systematically across the lunar near side, so that at one time or another, everything you can see on the Moon is close to the terminator. Depending on the time of the month, the terminator is either the place on the Moon where the Sun is rising, or the place where the Sun is setting. As you know from experience on Earth, shadows are longest when the Sun is rising or setting and they are shortest when the Sun is high in the sky, near noon. The length of the shadow when the Sun is at a known altitude tells the height of a lunar feature that casts it.

Giant impact: A theory of the origin of the Moon

Scientists know a lot about the ages of the rock in different terrains and parts of the Moon. They acquired these data using radioactive dating of samples from the hundreds of pounds of lunar rocks that were brought back to Earth by the six crews of NASA Apollo astronauts who landed on the Moon at different times from 1969 through 1972.

The Moon is a hostile place

When the Sun is up, the temperature on the lunar surface goes up to as much as 243° F (117° C), but at night it drops to around −272° F (−169° C). These extreme temperature changes are due to the absence of any meaningful atmosphere to insulate the surface and reduce the heat lost at night. There's no liquid water on the Moon. It's too hot, too cold, and too dry to sustain life as we know it on the Moon, even if there were air to breathe.

Before the Apollo Moon missions, several top experts confidently predicted that the Moon would be the Rosetta stone of the solar system. With no liquid water to erode the surface, no atmosphere worth mentioning, and no active volcanism, the surface of the Moon should include much primordial material from the birth of the Moon and the planets. That's what they thought. But the Apollo lunar samples threw rocks on their theory.

When a rock melts and then cools and crystallizes again, all its radioactive clocks are reset. Radioactive isotopes begin producing fresh daughter isotopes that are trapped in the newly formed mineral crystals. The Apollo Moon rocks show that essentially the whole Moon, or at least its crust down to a considerable depth, was melted well after 4.6 billion years ago. The very oldest surface rocks of the Moon are, at most, "only" 4.5 billion years old. The difference between 4.6 and 4.5 billion years is 100 million years. And, unlike the minerals in Earth rocks, which contain water bound up into the mineral structures, the Moon rocks are bone dry.

The theory that has emerged to explain all this evidence, and to avoid the objections that were posed to previous theories, is the *Giant Impact* theory of the origin of the Moon. According to this theory, the Moon is composed of material blasted out of the mantle of Earth by a huge object, with up to three times the mass of Mars, that struck the young Earth a glancing blow. And the Moon also is thought to contain some rock from the mantle of that long-vanished impacting object.

The Giant Impact on the young Earth knocked all this material up into space as a vapor of hot rock. It condensed and solidified like snowflakes. The snowflakes knocked into each other and stuck together, and before you know it, the Moon had formed. Coming together in powerful impacts of the last big pieces of accumulated rock, the Moon was melted by heat from the impacts.

All the impacts that caused the craters that we now see on the Moon happened later, although most of them date back to more than 3 billion years ago.

The Moon is less dense than Earth as a whole, and about as dense as Earth's mantle, according to this theory, because it was made from mantle material. This theory predicts that the Moon shouldn't have much of an iron core, if any. And a small core, in a small object (meaning the Moon), would have cooled and frozen long ago if it ever contained liquid iron. So the Moon shouldn't be able to generate a global magnetic field. And that's exactly what space measurements tell us. The Lunar Prospector, a satellite put in orbit around the Moon in the late 1990s, detected magnetic fields, but only at isolated places on the Moon. They are simply fossil magnetic fields, produced in an unknown way, long ago.

That's our best theory. Unfortunately, no one can think of a basic test for the Giant Impact theory of the origin of the Moon. There is no special kind of rock that is predicted by the theory, for example.

So it's a good theory, but we may never really know what made the Moon.

Enjoy the northern lights

The aurora is one of the most beautiful sights of the night sky, and for many people, a rare one. Depending on whether you live in the Northern Hemisphere or the Southern Hemisphere, you can see the *aurora borealis* (northern lights) or the *aurora australis* (southern lights), respectively.

The aurora is an eerie glow in the dark night sky that may remain stationary for minutes to hours (making it hard for a beginning observer to identify it) or constantly change. It can shimmer, pulsate, or even flash around the sky. The aurora may appear to you in many forms; here are a few of the most common:

✔ *Glow:* The simplest form of auroral display. The glow resembles a part of the sky where thin cloud is reflecting moonlight or city lights. But there's no cloud, just the eerie light of an aurora.

✔ *Arc:* Shaped like a rainbow, but with no sunlight to produce one. A steady or pulsating green arc is most common. But sometimes faint red arcs appear.

✔ *Curtain:* Also called *drapery.* This spectacular auroral form resembles a billowing curtain at a theatre, but nature is the star of this show.

✔ *Rays:* One or more long, thin bright lines in the sky, appearing like faint beams from the heavens.

✔ *Corona:* High overhead, a crown in the sky, with rays emanating in every direction.

Aurora are produced when streams of electrons from Earth's magnetosphere rain down on the atmosphere below, stimulating oxygen and other atoms to shine.

Aurora occur constantly in two geographical bands around Earth at high northern and southern latitudes. Folks who live beneath these two *auroral ovals* see aurora every night. But there are big exceptions: When a great disturbance in the solar wind strikes the magnetosphere, the ovals move equatorward. Then people in the *auroral zones* (the lands beneath the ovals) may miss their aurora, but those of us who rarely see them are treated to a great show. The most likely times are a few years after the peak of the sunspot cycle. So keep your eyes open for aurora, especially during the period from 2002 to 2003.

Check out the daily appearance of the auroral ovals with views and data from NASA, NOAA, and U.S. Air Force satellites at `solar.uleth.ca/solar/www/aurora.html`, and check the predictions of coming aurora at `solar.uleth.ca/monitor`, thanks to the good folks at Canada's University of Lethbridge.

Chapter 6

Earth's Near Neighbors: Mercury, Venus, and Mars

In This Chapter

▶ Probing planetary surfaces, atmospheres, and interiors

▶ Finding and observing Mercury, Venus, and Mars

▶ Understanding why Earth is different

You can spot Earth's neighboring terrestrial planets Mercury, Venus, and Mars with the naked eye and inspect them with your telescope. They are tantalizing objects, revealing only a little of their nature when viewed from Earth. That's why most of what scientists know about their physical properties, geologic forms, and likely histories is based on images and measurement data sent back to Earth by interplanetary spacecraft.

Mercury has been visited by a single spacecraft, which flew past it three times and went on into space. Venus has been visited, orbited, and even landed on by several probes. Mars has been the target of a great many probes, landers, and a robot rover; more are being sent every two years. The mapping of Venus and Mars is very thorough, but large parts of Mercury still have not been seen.

Hot, Shrunken, and Battered: Mercury's a Great Ball of Iron

Despite three passes by the Mariner 10 spacecraft in 1973 and 1974, less than half of Mercury has been mapped. The remainder either was not in view from Mariner 10, or was in darkness when it came by. To remedy this deficiency, NASA plans to launch a new probe to Mercury in 2004.

You can follow the progress in developing and launching this spacecraft, called MESSENGER (MErcury Surface, Space ENvironment, GEochemistry and Ranging) at the site `sd-www.jhuapl.edu/MESSENGER`. To view a video of the Mariner 10 images of Mercury, click MESSENGER Science, then click MESSENGER Details, and then click the words "Click here or on the Mariner 10 mosaic on the right for video animation of the first of the two Mercury flybys (8.8MB)."

Here's what scientists know so far, gathered mostly from Mariner 10 and from observations by radar astronomers on the ground, who transmit pulses of radio waves toward Mercury and study the echoes. Mercury's surface is like that of Earth's Moon, with one impact crater after the other. But Mercury has long winding ridges that cut across impact craters and other geologic features. They were probably caused by shrinkage of the crust, which was cooling from a molten state. And Mercury has fewer small craters than the Moon, in proportion to the number of large craters.

Highly cratered highlands are present on Mercury, as on Earth's Moon (Mercury has no known moon of its own). But unlike the Moon, Mercury's highlands are interrupted by gently rolling plains. Elsewhere, flat plains make up the Mercury lowlands.

The largest trace of any impact on Mercury is the Caloris basin. It's not fully mapped because much of it was in darkness when Mariner 10 came by. Astronomers' best estimates suggest that Caloris is about 830 miles (1,340 kilometers) across, which makes it among the largest impact basins in the solar system. Impact basins are huge craters like those lava-filled structures called maria on the Moon. On the *antipode* of Caloris, which is the spot opposite Caloris on Mercury, is a strange region of broken hills and valleys. The collision that caused the Caloris basin generated powerful seismic waves, which traveled through Mercury and around its surface, converging at the antipode with catastrophic effect.

Mercury has a density 5.4 times that of water. This high density means that Mercury has a huge iron core that constitutes the bulk of the planet. The outer layer of rock, called the mantle, must be no more than 380 miles (610 kilometers) thick. The presence of a global magnetic field, detected around Mercury by Mariner 10, suggests to many experts that some of that huge iron core must still be molten, although simple calculations indicate that the core should have cooled enough to solidify by now.

Faint traces of atmospheric gases exist on Mercury, but for practical purposes, it's airless like Earth's Moon. Extraordinary changes in temperature occur; temperatures can reach as much as 870° F during the day, and as low as –300° F at night. Areas of unusually high radar reflectivity near the North and South Poles may mean that there is actually a large amount of ice at the poles, in perpetually shadowed crater bottoms. MESSENGER will investigate whether this interpretation is correct.

Venus: Not a Nice Place to Live, or Even Visit

There's never a clear day on Venus; it's perpetually covered from equator to pole by a 9-mile (15-kilometer) thick layer of clouds of concentrated sulfuric acid. And there's no relief from the heat. Venus is the hottest planet in the solar system, with a surface temperature of 870° F that stays about the same from equator to pole, day and night.

If you think that the heat is bad, wait until you feel the barometric pressure: It's about 93 times the pressure at sea level on Earth. But forget about seas; there's no water on Venus. You can complain about the heat, but not the humidity — it's a dry heat, like they have in Arizona.

The bad news about the weather on the surface of Venus is that a perpetual rain of sulfuric acid falls all over the planet. The good news is that this rain is a *virga,* meaning rain that evaporates before it hits the ground.

Almost all the excellent images of the surface of Venus that you can find on NASA (and other) Web sites are not photographs at all. They are detailed radar maps, notably from the Magellan spacecraft. There aren't many photos of Venus's surface because the clouds block the view of telescopes on Earth and of any camera on a Venus-orbiting satellite. The cloud tops are at an altitude of 40 miles (65 kilometers), much lower than where a satellite can operate.

You can see satellite radar maps and lander images of Venus and more at the Views of the Solar System Web site at www.hawastsoc.org/solar/eng/homepage.htm. Click on Venus and then on Venus Photo/Animation Gallery.

The few images we have from Venus lander spacecraft, as pioneered by the former Soviet Union, show areas of flat rock plates, separated by small amounts of soil. They look like some areas of hardened basalt lava flows on Earth. But on Venus, things look orange at the surface, because the thick cloud cover has filtered the sunlight.

Flat plains, volcanic lowlands with rilles (the winding canyons left by lava flows) cover the vast majority of Venus (about 85 percent). This territory includes the longest known rille in the solar system, Baltis Vallis, which stretches across Venus for about 4,230 miles (6,800 kilometers). Cratered highlands and deformed plateaus are also present on Venus.

There aren't as many craters on Venus as you would expect, based on the number of them on Earth's Moon (Venus has no known moon of its own) and on Mercury. No small craters exist. There aren't many large craters because the surface of Venus was flooded with lava or reworked by volcanism (which

I define in Chapter 5) after its bombardment by impacting objects had mostly ended. This flooding or reworking erased all or most of the early craters. Few large objects have struck Venus since the early craters were destroyed, and small objects don't make many craters on Venus. Here's why: Objects capable of making craters up to two miles in diameter are impeded and destroyed by aerodynamic forces in the thick Venus atmosphere.

Huge volcanoes and mountain ranges are found on Venus, but nothing resembling the mountains on Earth that are caused by one crustal plate pushing into another. And Venus has no chains of volcanoes (like the Pacific "ring of fire"), which occur at the edges of plates. Plate tectonics and continental drift, as occur on Earth, don't take place on Venus.

Mars: A Planet of Mysteries

Mars has been topographically mapped with high accuracy, and you can find the latest chart of the entire planet at the NASA Web site (`ltpwww.gsfc. nasa.gov/tharsis/global_paper.html`). Topographically means that the altitudes of the landforms have been measured. The map comes from an instrument called a laser altimeter on the Mars Global Surveyor (MGS), a satellite in orbit around Mars. There's a camera on board MGS, too, and you can find its latest pictures at `www.msss.com`, the Web site of Malin Space Systems, a company that built and operates the camera.

Where has all the water gone?

The topographic map of Mars shows that most of the Northern Hemisphere is much lower than the Southern Hemisphere. The huge northern lowland may be the bed of an ancient sea, but even if it isn't, strong evidence suggests that liquid water was once common on Mars.

Mars is cold and dry now, with a great deal of ice at the poles. By one estimate, enough ice is present to flood the entire planet to a depth of 100 feet if it were melted. But the ice won't melt; Mars is just too cold. The atmosphere is mostly carbon dioxide, and in winter, some of that gas freezes on the surface, leaving thin deposits of dry ice. At the pole where winter is under way, a thin cap of dry ice often tops the permanent cap of water ice. Dry riverbeds with streamlined islands, and pebbles that look like they've been rounded in a torrent, are among the other evidence for past liquid water on Mars. The pebbles were imaged with the Mars Pathfinder (which landed on Mars) and its little robot, Sojourner.

A magnetometer on MGS discovered long parallel stripes of oppositely directed magnetic field frozen in the rocky crust of Mars. Mars doesn't have a global magnetic field today, but this finding may mean that it once had a global field that periodically reversed, just as Earth's field does (see Chapter 5), and it may also mean that Mars endured a crustal process resembling the seafloor spreading on Earth and producing a similar pattern. But Mars's molten iron core must have frozen solid long ago, so new magnetic field is no longer generated and the heat flow from the inside to the surface is so low that there's probably no volcanism still under way.

The volcanism that did occur on Mars produced immense volcanoes, such as Olympus Mons, which is about 370 miles (600 kilometers) wide and 15 miles (24 kilometers) high, or five times wider and almost three times higher than the largest volcano on Earth, Mauna Loa. Mars also has many canyons, including the immense Valles Marineris (Mariner Valley), 2,490 miles (4,000 kilometers) long. Impact craters appear, too. The craters are more worn down than those on Earth's Moon, because much more erosion has occurred, probably caused by the water that produced the great floods on Mars.

Mars has only two known moons, Phobos and Deimos. They're tiny, and are not visible with amateur telescopes.

Does Mars support life?

People have a lot of mistaken ideas about Mars, and some that might be right but have not been proven. These ideas all revolve around the possibility of life on Mars. Most of them are as improbable as the story about the future astronaut who returns from the planet. "Well is there life on Mars?" the reporters demand. "Not much during the week," he says, "but on Saturday night. . . ."

Claims for life strike out

The discovery of the "canals" on Mars spawned the first widespread speculation about the possibility of life. These were reported by some of the most famous astronomers of the late 19th and early 20th centuries. Planetary photography was not very useful in those days, because the exposures were fairly long and the images were blurred by atmospheric seeing (which I define in Chapter 3). So scientists believed that drawings by expert professional telescopic observers were the most accurate images of Mars. Some of these charts showed patterns of lines stretching and crisscrossing around the surface of Mars. Percival Lowell, an American astronomer, theorized that the straight lines were canals, engineered by an ancient civilization to conserve and transport water as Mars dried up. He concluded that the places where the lines crossed were oases.

Over the years, the idea of the "canals" and other reported indications of past or present life on Mars have struck out:

- ✔ When the American spacecraft Mariner 4 reached Mars in 1965, its photographs showed no canals, a conclusion verified in much greater detail, and for the whole surface of Mars, by images from subsequent Mars probes. That was strike one.

- ✔ Two later probes, the Viking Landers, conducted robotic chemical experiments on Mars to look for evidence of biologic processes such as photosynthesis or respiration. At first they appeared to have found evidence of biologic activity when water was added to a soil sample. But most scientists who reviewed the matter concluded that the water was reacting chemically with the soil, and there was no proof of life. This was strike two.

- ✔ Viking Orbiters also sent back images of the Mars surface as they revolved around the planet. These show, at one location, a crustal formation that — to some folks — looks like a face. Although many natural mountain peaks and stone formations on Earth resemble the profiles of the famous rulers, Native American tribal chiefs, and so on for whom they are named, some true believers claim that the "face on Mars" is a monument of some type, erected by an advanced civilization. Later, sharper images from MGS showed that this landform doesn't look like a face at all. That was "strike three" for the advocates of life on Mars.

But the idea of life was not "out," despite three strikes.

Fossil evidence?

In 1996, scientists analyzed samples of a meteorite that they believe comes from Mars, where it was knocked off the planet by the impact of a small asteroid or comet. The scientists found chemical compounds and tiny mineral structures that they interpreted as chemical by-products and possible fossil remains of ancient microscopic life. This work is very controversial, and most subsequent studies indicate that these conclusions are wrong. Based on current research, scientists can't make a persuasive case that supports the theory of past life on Mars.

The only thing to do is search systematically on Mars for evidence of life, past or present, in the regions that make the most sense — places where large quantities of water appear to have been present in the past, and where layers of sediment were deposited in ancient lakes or seas. Those are the types of places where people find the most fossils on Earth.

Observing the Terrestrial Planets

You can spot Mercury, Venus, and Mars in the night sky with the aid of monthly viewing tips from astronomy magazines and their Web sites (see Chapter 2) or with the aid of a desktop planetarium program. Venus is especially easy to find, because it's the brightest celestial object in the night sky, after the Moon.

Mercury orbits closest to the Sun, and Venus is next. They are both inside the orbit of Earth, so Mercury and Venus are always in the same region of the sky as the Sun, seen from Earth. Therefore, you can find these planets in the western sky after sunset or in the eastern sky before dawn. At those times, the Sun isn't very far below the horizon, so you can see objects that are near the Sun but to the west of it in the morning before the Sun rises, and you can see objects near but east of the Sun in the evening when the Sun sets. Your motto as a Mercury or Venus spotter should be to "Look east, young person" or, "Look west," depending on whether you're looking at dawn or dusk.

A bright planet appearing in the east before dawn is commonly called a *morning star,* and a bright planet in the west after sunset is an *evening star.* As Mercury and Venus move swiftly around the Sun, this week's morning star may be the same object as next month's evening star (see Figure 6-1).

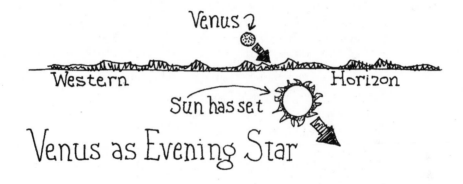

Venus as Evening Star

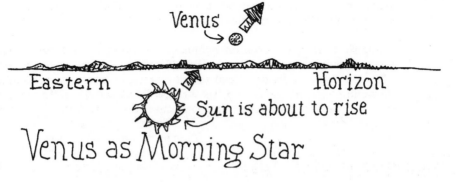

Figure 6-1: Venus can be a morning or an evening star, though it's not a star at all.

Venus as Morning Star

In the following sections, I explain the best time to observe these planets based on elongation, conjunction, and opposition — three terms that describe these planets' positions in relation to the Sun and the Earth — and how to use this understanding in your observations of the terrestrial planets.

Understanding elongation, opposition, and conjunction

Elongation, opposition, and conjunction are terms that describe a planet's position in relation to the Sun and the Earth. You will run across these terms when you check listings of the planets' positions in order to plan your observations. Here's what they mean:

- *Elongation* is the angular separation between a planet and the Sun, as visible from Earth. Mercury's orbit is so small that the planet never gets more than 28 degrees from the Sun. During some periods, it doesn't get farther than 18 degrees from the Sun, making it hard to spot. Venus can get up to 47 degrees from the Sun.

- *Greatest western (or eastern) elongation* occurs when a planet is as far from the Sun as it will get during a given apparition (when the planet is visible from Earth). Some greatest elongations are greater than others, because sometimes Earth is closer to the planet than at other occasions. Elongation is especially important when you are observing Mercury, because the planet is usually so close to the Sun that the sky at its position is not very dark.

- *Opposition* occurs when a planet is on the opposite side of Earth from the Sun. This never happens for Mercury or Venus, but Mars is at opposition about once every 26 months. That's the best time to observe it, because it appears largest in a telescope. And, at opposition, Mars is highest in the sky at midnight, so it's up and can be viewed all night.

- The term *conjunction* is often used when two solar system objects are near each other in the sky, as when the Moon passes near Venus, as we see them. In fact, Venus is far beyond the Moon, but that's a conjunction of the Moon and Venus.

Conjunction also has a technical meaning as well. Instead of describing positions in right ascension (the position of a star measured in the east-west direction) and declination (the position of a star measured in the north-south direction), astronomers sometimes use ecliptic latitude and longitude. The ecliptic is a circle in the sky that represents the path of the Sun through the constellations. *Ecliptic latitude* and *longitude* measure degrees north and south (latitude) or east and west (longitude) with respect to the ecliptic. (Don't worry, you won't need to use the ecliptic system. But knowing about it helps you understand the definitions that follow.)

Identifying superior and inferior conjunctions

Suppose that a planet is outside Earth's orbit. And it's at the same longitude as the Sun, so it's in the same east-west direction. That planet is at *superior conjunction* (see Figure 6-2).

Superior conjunction is a bad time to observe the planet, because it's on the far side of the Sun. So don't try to observe Mars at superior conjunction; you won't see it. The best time to observe Mars is at opposition.

However, Mercury and Venus are inside Earth's orbit. These two planets can be at the same longitude as the Sun when they are between Earth and the Sun, in *inferior conjunction*, as shown in Figure 6-3, or when they are on the far side of the Sun in superior conjunction.

You can see Venus best at inferior conjunction, when it looks largest and brightest. But Mercury is too close to the Sun for viewing at inferior conjunction; it's best viewed at greatest elongation.

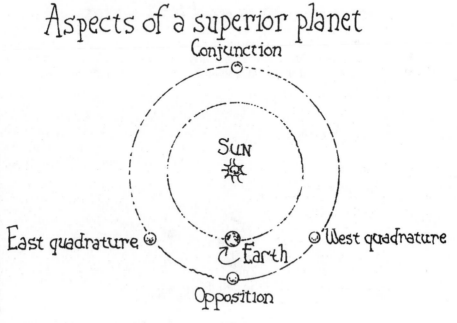

Figure 6-2:
A planet in superior conjunction is in the same east-west direction as the Sun.

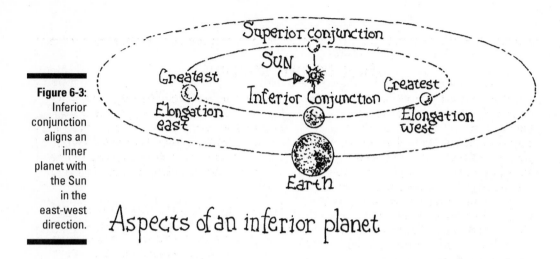

Figure 6-3:
Inferior
conjunction
aligns an
inner
planet with
the Sun
in the
east-west
direction.

In the sections that follow, I introduce you to observing the terrestrial planets in the order of ease, starting with Venus, which is the easiest.

Viewing Venus and its phases

The simplest planet to find is Venus. Venus is so bright that people with no knowledge of astronomy frequently notice it and call radio stations, newspapers, and planetariums to ask what "that bright star" is.

When scattered clouds move from west to east in front of Venus, the scene is often misunderstood. Folks think that it's Venus (which they don't recognize) that's moving rapidly, in the opposite direction of the clouds. Due to its brightness and the mistaken impression that it's moving rapidly behind a cloud deck, Venus is often reported as an unidentified flying object. It's not. We know it well.

After you become familiar with Venus, you may find that you can even spot it in broad daylight. Often, Venus is bright enough that if the sky is clear, with little haze or smog, you may spot it in the daytime by using *averted vision.* That means that you can glimpse it "out of the corner of your eye." For some reason, it's often easier to spot a celestial object with averted vision than if you look right at it. This may be a survival trait; it makes it a little harder for an enemy or a predator to sneak up on you from the side. But whatever the cause, it's a good thing to know about when you observe the skies.

A small telescope will show you Venus's most recognizable features: its phases and changes in apparent size. Venus has phases like the phases of Earth's Moon, and for the same reason: Sometimes part of the Venus hemisphere that is facing the Sun (and is therefore bright) is directed away from Earth. So a telescopic view of Venus shows a disk that is partly illuminated and partly dark.

Just an arc second

Apparent sizes in the sky are measured in angular units: Something that goes all the way around the sky, such as the Celestial Equator, is 360 degrees long. The Sun and the Moon, in comparison, are each about a half-degree across. The planets are much smaller, so smaller units are necessary to describe them. A degree is divided into 60 minutes of arc, and a minute of arc — called an arc minute or arc min — is divided into 60 seconds of arc — usually abbreviated as arc seconds or arc sec. There are 3,600 (60 times 60) seconds of arc in a degree. In many astronomy books and articles, a minute of arc is represented by a single prime symbol ('), and a second of arc is represented by a double prime (''). These are often mistaken by the uninitiated as the abbreviations for feet and inches. You can tell when a clueless copy editor has had the last cut at an astronomy article, because you will find a statement like "The Moon is about 30 feet in diameter."

Venus is actually only about 5 percent smaller in diameter than Earth. Its apparent size or angular diameter ranges from about 10 arc sec, when Venus is furthest away (and with a full Moon shape), to about 58 arc sec in diameter when it's closest and is a narrow crescent.

The dividing line between the bright and dark parts of Venus is called the *terminator*, just as on the Moon. But don't confuse it with the Arnold Schwarzenegger character. This terminator is perfectly safe; it's just an imaginary line on Venus.

As Venus and Earth orbit the Sun, the distance between the two planets shifts substantially. At its closest to Earth, Venus is a mere 25 million miles away, and at its furthest, it's a full 160 million miles distant. What's important here is the proportional change: At closest approach, Venus is about six times nearer to Earth than at its most distant position. And it looks six times bigger through a telescope.

What you won't see when you view Venus are striking features, like the Man in the Moon. Venus is totally covered by thick clouds, and all you can see is the top of the clouds. Venus is so bright because it's relatively close to both the Sun and Earth, and because it has a nice, bright reflecting cloud layer. But sometimes you may be able to discern the horns of the Venus crescent extending further into the dark side than predicted for the phase on that day. What you are seeing is some sunlight that has bounced around in the atmosphere of Venus and passed beyond the terminator into the side of the planet where it is night.

Images of Venus with striking cloud patterns, like those you see in books, were made in ultraviolet light, where the patterns show up. Ultraviolet light doesn't pass through our atmosphere (let's hear it for the ozone layer, which blocks this hazardous radiation), so you can't view Venus in it. In fact, you can't see ultraviolet light anyway; it's invisible to the human eye. But tele-scopes on satellites and space probes above or beyond the atmosphere can take ultraviolet pictures.

On rare occasions, observers report a pale glow on the dark part of Venus. This alleged glow, called the *ashen light,* is sometimes a real phenomenon and sometimes a trick of the imagination. After centuries of study, experts still can't explain the ashen light, so some of them deny that it exists. But with luck, you might see it. People claim to see other phenomena on Venus through their telescopes, but almost all of these reports are wrong. Experiments show that the effect is psychological: If people view a feature-less white globe from a distance, they may discern patterns that don't exist.

Watching Mars as it loops around

Mars is a bright red object, but not nearly as dazzling as Venus. So you have to check your sky maps to make sure that you don't mistake a bright red star, such as Antares in Scorpius (whose name means "rival of Mars") for the red planet.

The great advantage in observing Mars is that when it's in the night sky, it's often visible for much of the night. It's not like Mercury and Venus, which set fairly soon after sunset or rise only shortly before dawn. You usually have time for dinner and the nightly news before you head into the backyard to check on Mars.

With a small telescope, you can spot at least a few dark markings on Mars. The best times to see them last a few months, but occur only about every 26 months, when Mars is at opposition. At opposition, Mars looks biggest and brightest, and it's easiest to see details on its surface.

EYES ON THE SKIES

Venus in transit

One of the rarest planetary events you can see is a *transit* of Venus, when the planet passes right in front of the Sun and appears as a tiny black disk against the bright solar surface.

This event can be seen with the naked eye (be sure to use a safe solar filter, as I describe for sunspot observing in Chapter 10!), but no living astronomer has seen one. That's because the last transit of Venus was in 1882.

You can observe this rare phenomenon, how-ever, because two transits are coming up soon, on June 8, 2004, and June 6, 2012. You may have to take a trip to see them, but the effort is worth it to see such an unusual event.

EYES ON THE SKIES

Leaping backward or forging ahead?

A basic project for beginning planet gazers is to track the motion of Mars across the constellations; all you need are your eyes and a sky map.

Locate Mars among the stars and mark that position with soft pencil on your map. Repeat this observation on each clear night, and you can see a pattern emerge that puzzled the ancient Greeks and led to complicated theories, most of them wrong.

Most of the time, Mars moves eastward from night to night, just as Earth's Moon moves eastward across the constellations. The Moon keeps going, but Mars sometimes reverses course. For two to almost three months (62 to 81 days) at a time, Mars heads west across the constellations,

moving backward for 10 to 20 degrees. Then it gets back on track and heads east again. The backtracking is called Mars's *retrograde motion.*

This backtracking is not a case of Mars not knowing whether it's coming or going. The retrograde motion is just an effect produced by Earth racing around the Sun. While you chart Mars's motion, you are standing on Earth, which is racing around the Sun once every 365 days. Mars is going slower, making one full orbit in 687 days. As a result, when we pass Mars on our inside track (lapping it, as it were), Mars seems to move backward against the reference frame of the distant stars. But in reality, Mars is always forging ahead.

The upcoming oppositions of Mars are in

> June 2001
>
> August 2003
>
> November 2005
>
> December 2007
>
> January 2010

Don't miss them!

The very best Mars opposition of the decade will be on August 27, 2003, when Mars will have an apparent diameter of 25 arc sec, or half-again as large as at the opposition in April 1999. At its best oppositions, Mars is south of the Celestial Equator, but can still be observed well from temperate latitudes in the Northern Hemisphere.

The easiest Martian surface feature to spot with a small telescope is usually Syrtis Major, a large dark area extending northward from the equator. Mars's day is nearly the same as Earth's: It's 24 hours, 37 minutes. So if you look at Mars off and on during a night, you may be able to see Syrtis Major move slowly across the planet's disk as Mars turns. Experienced amateur planetary observers may see its polar caps and other markings as well.

The key to planetary observing is to take advantage of moments of *good seeing,* which means that the atmosphere above your telescope is steady. At those times, the stars don't twinkle quite as much, and you can use a higher magnification eyepiece with the telescope to bring out fine detail on Mars or another planet. When the seeing isn't very good, the telescopic image is blurred and seems to jump around. Under those conditions, high magnification is useless; it just magnifies the blurred, jumping image and you are better off with a low-power eyepiece.

NASA images of Mars, taken by interplanetary probes and the Hubble Space Telescope, are much too detailed to guide you in small-telescope observation. You need a simple *albedo map,* which charts and names the bright and dark areas on Mars as visible with small telescopes. Even an albedo map offers more detail than the average observer ever sees, but it's a good guide and a challenge to your observing skills. You can find such a map in *Norton's Star Atlas and Reference Handbook (Epoch 2000.0),* 19th Edition — by Arthur P. Norton, Ian Ridpath (Editor), published by Longman Publishing Group, 1998 — or at `mpfwww.jpl.nasa.gov/mpf/marswatch/marsnom.html`, the Mars Watch site.

Astronomers rate sky conditions in terms of *seeing* (the steadiness of the atmosphere above the telescope), *transparency* (the freedom from clouds and haze), and *sky darkness* (freedom from interfering artificial light, moonlight, or sunlight). When observing a bright planet such as Mars, good seeing is the most important, and dark sky is least important. But the darker the sky, the steadier the air, and the higher the transparency, the better you will enjoy the night.

Unfortunately, even when atmospheric conditions are ideal at your observing site and when an opposition of Mars is in progress, disaster may strike. Mars is a planet that experiences worldwide dust storms, which hide its surface features from view.

In fact, professional astronomers are asking amateur astronomers to help monitor Mars, to let them know when a dust storm begins and to report other pronounced changes in the appearance of the planet. You can get information on this program at Cornell University's MarsNet site, `astrosun.tn.cornell.edu/marsnet/mnhome.html`. It's a lot more fun to get a good sharp view of Mars, but if all else fails, at least you may get credit for discovering a dust storm. The experts will welcome your dust report, not brush it off.

You need experience to become a reliable telescopic Mars observer. When you are a beginning observer, just because you can't make out any detail, don't assume that a great dust storm is in progress. Get accustomed to seeing Mars in detail. Only then should you consider that, when you can't see details, it's the planet that's at fault and not your inexperience.

There's a famous motto in science, "The absence of evidence is not necessarily evidence of absence." You may not see detail the first time you look, but that doesn't mean that a dust storm is obscuring your view. As a telescopic observer, you have to train your viewing skills, just as gourmets and wine lovers train their palates.

Outdoing Copernicus by observing Mercury

It's said that the great 17th-century Polish astronomer Nicholas Copernicus, who proposed the *heliocentric* (Sun-centered) *theory* of the solar system, never spotted the planet Mercury.

But Copernicus didn't have modern aids, such as desktop planetariums, astronomy Web sites, and monthly astronomy magazines. You can use those aids to find out when Mercury will be best placed for observation during the year. Those are the times of greatest western and eastern elongation, which occur about six times each year.

At temperate latitudes, such as those of the continental United States, Mercury is usually only visible in twilight. By the time the sky is dark, well after sunset, Mercury has set, too. And in the morning, you won't spot Mercury until the impending dawn is starting to light the sky. It will resemble a bright star, but appear much dimmer than Venus in the west at dusk or in the east before dawn.

For Mercury, be an early riser

Mercury is much smaller than Venus, but you can see its phases through your telescope. The best time to do this is when Mercury is at western elongation and appears in morning twilight. The atmospheric steadiness or seeing is almost always better low in the east near dawn than it is low in the west after sunset. So you will get a much sharper view in the morning. Standard guides like the highly regarded *Observer's Handbook* (an annual publication of The Royal Astronomical Society of Canada, www.rasc.ca), the *Astronomical Calendar* (published annually by Universal Workshop, www.kalend.com), and the astronomy magazines and their Web sites all tell when the elongations are due.

You need a viewing site with a clear eastern horizon, because Mercury doesn't get very high in the sky when the Sun is below the horizon. If you have trouble spotting it with your naked eye, sweep around that part of the sky with a pair of low-power binoculars. And if you have a computerized telescope with a built-in database, you can just punch in "Mercury" and let the telescope do the finding.

Mercury in transit

Like Venus, Mercury is sometimes seen in transit, when it passes across the face of the Sun and appears as a small black disk against the solar surface, as visible from Earth. Observe a transit of Mercury with a telescope, using the procedures for safe solar viewing that I explain in Chapter 10. (Remember, you're viewing Mercury against the Sun, so solar viewing precautions are absolutely necessary). Two transits of Mercury will occur in the next decade, on May 7, 2003, and November 8, 2006. Depending on where you live, you will probably have to travel to see one of them.

Don't expect to see surface markings

Seeing surface markings on Mercury with a small telescope, or indeed with almost any telescope on Earth, is extremely difficult. Mercury's apparent size at greatest elongation is only about 6 to 8 arc sec.

Some experienced amateur observers report seeing surface markings, but no useful information has ever come from such sightings. A few of the greatest planetary observers of all time thought that they could see and draw the surface markings. From these drawings, these observers tried to deduce the rotation period or "day" of Mercury. These experts found that the day was equal to the of Mercury, 88 Earth days. But they were wrong. Radar measurements later proved that Mercury turns once every 59 Earth days.

Nevertheless, once you learn to spot Mercury by eye and then check on its phases with your telescope, you will be way ahead of Copernicus!

You can get more information on observing Mercury and the other planets from the Association of Lunar and Planetary Observers (ALPO), which also collects sketches and other planetary observations made by amateur observers, and offers observing forms, charts, and other publications. Some of their advice is more optimistic than mine concerning what is visible in small telescopes, but why not try (to paraphrase a slogan of the U.S. Army) to "See all that you can see." The ALPO home page is at www.lpl.arizona.edu:80/alpo.

TIP

Why Mercury lovers choose morning

Here's why the seeing is better near the dawn horizon than near the sunset horizon: By sunset, the Sun has warmed the surface of Earth all day, so as you look out low in the sky to the west, you are looking through turbulent currents of warm air rising from the surface. But in the morning, Earth has had all night to cool down and stabilize. It takes a few hours for the rising Sun to warm the land and mess up the seeing again.

Discovering Why Earth Is Best: Comparative Planetology

Mercury is a tiny world of extreme temperatures, but it has a global magnetic field like Earth's, implying the presence of a molten iron core like Earth's. Although Venus and Mars don't have global magnetic fields, they are similar to Earth in many other ways. But liquid water and obvious, abundant life occur today only on Earth. What makes Earth different?

Venus, unlike Earth, has a hellish temperature. Venus is farther from the Sun than Mercury, but is even hotter. The high temperature is due to an extreme greenhouse effect, the process by which atmospheric gases raise the temperature by absorbing outward flowing heat. Earth's atmosphere may once have contained large amounts of carbon dioxide, the way the atmosphere of Venus does now. But on Earth, the oceans absorbed much of the carbon dioxide, and that gas couldn't trap the heat the way it does on Venus.

Mars, on the other hand, is too cold to support life. Mars has lost most of its original atmosphere. Its atmosphere isn't thick enough to produce a greenhouse effect sufficient to warm the surface above the freezing point of water.

The three large terrestrial planets are like the bowls of cereal in the child's story of Goldilocks. Venus is too hot, Mars is too cold, but Earth is *just right* for liquid water and life as we know it. Putting together the information on the basic properties of the terrestrial planets and their respective differences, we can conclude that

✔ Mercury is like the Moon on the outside but like Earth on the inside.

✔ Venus is Earth's "evil twin."

✔ Mars is the little Earth that died.

Earth is the Goldilocks planet — Just right!

Chapter 7

The Asteroid Belt and Near Earth Objects

In This Chapter

▶ Finding out where asteriods came from

▶ Evaluating Earth's risk of a dangerous asteroid impact

▶ Finding out what scientists are doing about the threat

▶ Observing asteroids

*A*steroids are big rocks that circle the Sun. The vast majority are safely beyond the orbit of Mars, but thousands of asteroids are in orbits that come close to or cross Earth's orbit. Many scientists believe that an asteroid hit Earth about 65 million years ago, wiping out the dinosaurs and many other species.

In this chapter, I introduce you to these big rocks and explain how to observe them. And, in case you're worried, I tell you the truth about the risk of an asteroid hitting Earth in the future, and fill you in on the research scientists are conducting to deal with the possibility.

Asteroids: Leftovers from the Birth of the Solar System

Asteroids are often called minor planets. Astronomers believe that they are remnants of the formation of the solar system — objects that never combined to grow into planets. Some asteroids, such as Ida, even have their own moons (see Figure 7-1).

Photo Courtesy of NASA

Figure 7-1:
The asteroid
Ida has its
own moon,
Dactyl.

Asteroids range in size from the largest, Ceres, which is 580 miles (933 kilometers) in diameter, to large meteoroids. (A boulder-sized space rock is a very small asteroid or a very large meteoroid; take your pick.)

As of 1999, about 10,000 asteroids were known to science, with dozens more being discovered every month or two. Accurate orbits have been determined for about 6,000 of them. You can readily see the largest asteroids, such as Ceres and Vesta, through small telescopes (see the later section, "Small Points of Light: Searching for Asteroids").

Ceres and Vesta are so big that their own gravity makes them round. But smaller asteroids are often potato-shaped, and frequently look like they have been blasted apart (see Figure 7-2). Indeed, they have. The asteroids in the belt are constantly bumping into each other and breaking off big and little chips. The big chips are simply smaller asteroids, and the little chips are asteroidal meteoroids.

Most of the known asteroids are in a huge, flat region centered on the Sun and located between the orbits of the planets Mars and Jupiter. This region is called the *Asteroid Belt*. At rare intervals, a small asteroid (or a large meteoroid, take your pick) smashes into Earth. One of these impacts caused the famous Meteor Crater (which ought to be called Meteoroid Crater, or Asteroid Crater) in northern Arizona, near Flagstaff. It's well worth a visit.

The Moon is covered with impact craters. On Earth, most impact craters have been eroded by the action of weather and geological processes, such as mountain building, erosion, and volcanism. You can see aerial photographs of many of Earth's best impact craters at the Views of the Solar System site at www.hawastsoc.org/solar/eng/tercrate.htm.

Asteroids are too small to show much in the way of surface features when seen from Earth, even with the most powerful telescopes; for the most part, they just look like stars. (A rare exception is Vesta. At the Keck Observatory Web site at www2.keck.hawaii.edu:3636/realpublic/ao/solarsys.html, you can see a movie of Vesta spinning as it flies through space.) But you can see asteroids moving against the starry background if you make telescopic observations an hour or two (or a night or two) apart.

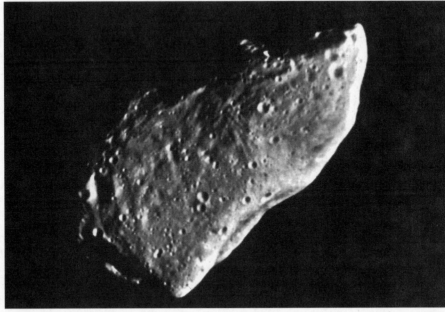

Figure 7-2:
Some
asteroids
resemble big
potatoes.

Photo Courtesy of NASA

Near Earth Objects: Is Earth Endangered?

Not all asteroids are located safely beyond Mars. Thousands of small aster-oids are in orbits that cross or come close to Earth's orbit. These are the logically named *Near Earth Objects (NEOs)*, and they include about 170 that have been classified as *Potentially Hazardous Asteroids (PHAs)*, meaning that some day their paths might take them uncomfortably close to Earth, or they might even strike our planet. The Minor Planet Center of the International Astronomical Union keeps tabs on PHAs, and several observatories are sweeping the skies to discover more of them.

Astronomers don't know of any specific object that is currently menacing Earth. Conspiracy theorists think that if astronomers did know about a doomsday asteroid, we wouldn't tell. But let's face it, if I knew the world was in danger, I would be settling my affairs and heading for the South Seas, not writing this book.

In 1998, the Hollywood movies *Armageddon* and *Deep Impact* gave sensational-ized versions of what might happen if a large asteroid or comet were on a collision course for Earth. Such catastrophe stories are inspired, in part, by the widely accepted conclusion that an asteroid about 6 miles (10 kilometers) wide struck Earth about 65 million years ago. The Chicxulub crater, a 110-mile (180-kilometer) wide geologic formation that is partly on Mexico's Yucatan peninsula and partly offshore in the Gulf of Mexico, may be the surviving trace of this impact, which is said to have wiped out the dinosaurs. (It certainly didn't do them any good.)

For a brief time in March 1998, many people feared that a small, newly discovered NEO might have the potential to strike Earth in the year 2028. That possibility was eliminated within a day when additional observations showed that the asteroid orbit would not intersect Earth. And some experts even disagreed with the initial prediction — as experts so often do.

Although Earth appears to be safe for now, an NEO that is on a collision course with Earth might be discovered in the future, so scientists are thinking about what could be done in such an eventuality.

When push comes to shove: Nudging an asteroid

Some experts propose that we develop a powerful nuclear missile to intercept a killer asteroid before it could strike. But if an asteroid were heading our way and we blew it up, the results could be worse than the impact of the intact asteroid. It would be like the scene in the Walt Disney movie *Fantasia,* in which the sorcerer's apprentice chops up the magic broom that has gotten out of control and won't stop fetching water. He just gets a whole bunch of little brooms that each start fetching water.

If we blew up an asteroid with a nuclear bomb, a swarm of smaller rocks — instead of one big rock heading for Earth — would be on that deadly trajectory, like the warheads on a multiple independently targeted reentry vehicle or MIRV. A MIRV is the military's most powerful ballistic missile weapon. It's launched with a bunch of nuclear bombs on board, each of which is released and aimed at a different enemy target. But those rocks would pack more deadly energy than all the Pentagon's weapons. So a better idea is to use the nuclear missile (or perhaps some other kind of missile) to only nudge the asteroid, so that it is a little early or a little late on its path. Then it would miss Earth. Phew!

The problem with nudging an asteroid is that scientists don't know how much force to apply. We don't want to break it up, but because the mechanical strength of asteroids is unknown, we don't know how hard to hit it. Asteroids may be made of strong rock or fragile stone. Some may be mostly solid metal. If you don't know your enemy, you could make things worse by striking him.

In *Fantasia,* the sorcerer himself broke the spell on the enchanted broom, but without a sorcerer to make the asteroids disappear, we need hard information to design a system that can safely protect Earth from asteroids.

Forewarned is forearmed: Surveying NEOs

Astronomers have a plan to help design a system that will protect Earth from renegade asteroids:

✔ First, take a census of the Near Earth Objects, to make sure that we have located every rock that's a kilometer or more in size (a mile is about 60 percent longer than a kilometer) in our bailiwick. Those are the ones that are so big and come so close that they pose the greatest potential menace.

✔ Then, track these NEOs and compute their orbits to determine if any are likely to strike Earth in the foreseeable future.

✔ Finally, study the physical properties of asteroids to learn as much as we can about them.

✔ Then, when we understand the threat, design a missile to counteract it.

To survey the NEOs, special-purpose asteroid discovery telescopes are in operation at several locations. You can visit their Web sites and see what they have been finding lately.

Two of the most important are

✔ The Lincoln Near Earth Asteroid Research (LINEAR) project at White Sands, New Mexico, funded by the United States Air Force, www.ll.mit.edu/LINEAR

✔ NASA's Near-Earth Asteroid Tracking (NEAT) project, with observations from a Hawaiian observatory, huey.jpl.nasa.gov/~spravdo/neat.html

A private organization, the Spaceguard Foundation, is dedicated to saving Earth from killer asteroids. They may have bitten off more than they can chew; just saving the whales or the spotted owls is hard enough. But you can look them up and join at the spaceguard.ias.rm.cnr.it/SGF Web site.

A list of Potentially Hazardous Asteroids is maintained by the Minor Planet Center at the cfa-www.harvard.edu/iau/lists/Dangerous.html Web site. Probably none of these asteroids is larger than about 10 miles in diameter, and most are smaller. But a rock a few miles in size, striking Earth at 25,000 miles per hour (11 kilometers per second), would be a far greater catastrophe than the simultaneous explosion of all the nuclear weapons ever made. That would be a rare case when astronomy isn't fun.

Small Points of Light: Searching for Asteroids

Looking for asteroids is like discovering comets (see Chapter 4), except that you are looking for a small spot of light that is not fuzzy, thus resembling a star. But unlike a star, the asteroid moves perceptibly against the background of other stars from hour to hour and from night to night.

You can easily see the largest asteroids, such as Ceres and Vesta, through small telescopes; charts to guide you are published in advance of good viewing periods in the astronomy magazines. Most good planetarium programs also make sky maps that show you where they are.

Table 7-1 lists the four biggest objects in the asteroid belt. The two largest, Ceres and Pallas, are at nearly the same average distance from the Sun, although Pallas's orbit is much more elliptical than Ceres's.

Table 7-1	The "Big Four" Asteroids		
Name	**Diameter in Miles**	**Diameter in Kilometers**	**Mean Distance from the Sun (A.U.)**
Ceres	580	934	2.77
Pallas	327	526	2.77
Vesta	317	510	2.36
Hygiea	254	408	3.14

Searches for currently unknown asteroids are often made by advanced amateurs using electronic cameras on their telescopes. They make a series of images of selected areas of the sky, generally in the direction opposite the Sun (which, of course, is below the horizon). When you see a small point of light (like a star) change its location from one image to the next, it's probably an asteroid.

You will not be ready to search systematically for unknown asteroids until you become a skilled amateur astronomer with a few years of experience. But as soon as you are adept at using a telescope, you can observe some of the well-known asteroids. When they are in view, you can find short articles and sky charts — check the astronomy magazines and sky-view Web sites — that will guide you to them.

Timing asteroidal occultations

An *occultation* is a kind of eclipse that occurs when a moving body in the solar system passes in front of a star. Occultations are caused by the Moon (lunar occultations), asteroids (asteroidal occultations), and planets (planetary occultations). They can also be caused by the moons and rings of planets and by comets.

You can enjoy an occultation without obtaining scientific data, but what a waste of a unique opportunity! The details of an occultation differ from place to place on Earth. From occultation data, astronomers can get a more accurate picture of a number of sky objects. For example, sometimes the occultations reveal that what seems to be an ordinary star is actually a close *binary system,* two stars in orbit around their common center of mass.

To make your observation scientifically useful, you need to time it accurately and to know the exact location (latitude, longitude, and altitude) where you observed it. In the past, observers figured out the location by consulting topographic maps. But nowadays, if you observe in a group, someone is very likely to have a GPS (Global Positioning System) receiver, such as those sold to boat operators and private pilots. They are widely available for less than $300, and they give an accurate readout of your precise location.

Helping to track an occultation

Occultations by asteroids are much trickier to observe than lunar occultations, because they often cannot be predicted with sufficient precision. Astronomers go to various spots on the predicted occultation ground track and attempt to observe it. But because the diameters, orbits, and shapes of most asteroids are not accurately known, the predictions can't be precise. Because the occultation may be visible at some places and not at others, volunteers to monitor an asteroidal occultation are needed at many locations.

Amateur observations like these help determine the sizes and shapes of the asteroids involved in the occultations.

You can get the latest predictions of occultations on the Web site of the International Occultation Timing Association (IOTA) at lunar-occultations. com/iota/iotandx.htm.

IOTA recommends that you begin occultation study by observing with an experienced astronomer, just to get the hang of it.

Chapter 8

Jupiter and Saturn: Great Balls of Gas

In This Chapter

▶ Understanding the recipe for gas-giant planets

▶ Focusing on Jupiter's brilliance

▶ Spotting the Great Red Spot

▶ Setting your sights on Jupiter's moons

▶ Visiting Saturn's rings and moons

*J*upiter and Saturn are among the best sights in a small telescope, and one or both of them is usually well placed for observations. The four largest moons of Jupiter and the famous rings of Saturn are favorite targets when amateur astronomers give friends and family some peeks through their telescopes. And the underlying science of these huge planets and their satellites is fascinating, too.

Inside Jupiter and Saturn: What You See Is Not What You Get

Jupiter and Saturn are like hot dogs with unapproved food coloring. The meat isn't the mystery, the additives are. What you see on Jupiter and Saturn are the clouds, made of ammonia ice, water ice (like the cirrus clouds on Earth), and a compound called ammonium hydrosulfide. Water-drop clouds may be a part of the mix, too. But those appearances are deceiving. The cloud materials are trace substances. Jupiter and Saturn are mostly hydrogen and helium, like the Sun. And, despite much theorizing, scientists have no idea what chemicals make the Great Red Spot on Jupiter red or produce any of the other off-white tints in the clouds of the two great planets.

Jupiter and Saturn are the two larger of the four gas-giant planets (the others are Uranus and Neptune). Jupiter has 318 times the mass of the Earth; Saturn surpasses the Earth's mass by about 95 times. As a result, their gravity is enormous, and inside them, the weight of the overlying layers produces enormous pressure. Descending into Jupiter or Saturn is like sinking in the deep sea. The further down you go, the higher the pressure. Don't even think about scuba diving there. The pressures are immense, and unlike the sea, the temperature increases radically with depth.

Up at the atmospheric levels where we can see, in the cloud decks, the temperatures drop to -236° F (-149 C) on Jupiter and -288° F (-178 C) on Saturn. But at great depths, the squeeze is on. By the time you've reached 6,200 miles (10,000 kilometers) below the clouds on Jupiter, the pressure has soared to one million times the barometric pressure at sea level on Earth. And the temperature equals that of the visible surface of the Sun! But Jupiter is weirder than the Sun. The density of the thick gas at this depth is much higher than that at the solar surface, and the hot hydrogen is compressed so that it behaves like a liquid metal.

Swirling currents of this liquid metal hydrogen generate powerful magnetic fields on Jupiter and Saturn that reach far out into space.

The Earth derives almost all its energy from the Sun, but Jupiter and Saturn glow intensely in infrared light, each generating almost as much energy as it gets from the Sun. The Earth's internal heat is from radioactivity, liberated from radioactive substances such as uranium. But the huge gravity of Jupiter and Saturn is compressing them, and when you compress a gas, you heat it. So deep inside, these planets are extremely hot. This upward moving heat, together with the downward shining rays of the Sun, stirs up their atmospheres and produces jet streams, hurricanes, and other kinds of atmospheric storms that continually change the appearance of these planets.

Observe Jupiter, by Jove!

Jupiter has about 1,000th the mass of the Sun. Sometimes it's called "the star that failed." If it were only 80 or 90 times more massive, the temperature and pressure at its center would be so high that nuclear fusion would begin and keep going. Then Jupiter really would be a star!

Jupiter is easy to find, because, like Venus, it is brighter than any star in the sky. (A small exception: When it's on the far side of the sun, Jupiter may be slightly fainter than the brightest star, Sirius.) If you have a computer-controlled telescope that can point to the position of the planet, or you know *just* where to look, you can sometimes see Jupiter in the daytime.

Jupiter is really a huge ball of gas that stretches 44,432 miles (71,492 kilometers) across at its equator. The great planet rotates at enormous speed,

Figure 8-1:
Jupiter and
its band of
clouds.

Photo Courtesy of NASA

making one complete turn in only 9 hours, 55 minutes, 30 seconds. The rapid spin helps produce ever-changing bands of clouds, parallel to Jupiter's equator. What you see in your telescope when you view Jupiter is really the top of the planet's clouds. Depending on the viewing conditions, the size and quality of your telescope, and circumstances on Jupiter itself, you may see from as few as 1 to as many as 20 cloud bands (see Figure 8-1).

Jupiter's darker bands of clouds are called belts; the lighter ones are zones. Right down the center of the disk is the Equatorial Zone, flanked by the North and South Equatorial Belts (NEB and SEB). In the SEB is the Great Red Spot, often the most conspicuous feature on Jupiter. This atmospheric disturbance, sometimes compared to a great hurricane, has hovered in the Jupiter atmosphere for at least 120 years. In fact, the Great Red Spot may have been seen as early as 1664, although if so, it faded away before reappearing in the 19th century.

In search of the Great Red Spot

The Great Red Spot, shown in Figure 8-2, is a storm as big as the Earth and sometimes bigger. Like most of Jupiter's features, it may change from day to day. Its color can grow paler or deeper. White clouds that are big enough to see using some amateur telescopes form near the spot and move along the South Equatorial Belt. Sometimes a cloud in the SEB or another belt seems to be drawn out across the planet, stretched mostly in longitude. A cloud with that shape is known as a *festoon,* and spotting this interesting display is indeed a festive occasion!

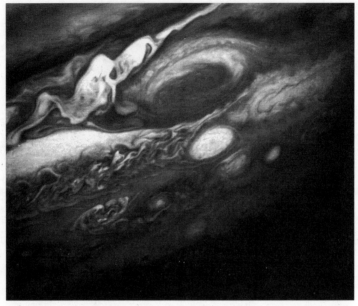

Figure 8-2:
Jupiter's
Great Red
Spot makes
for stormy
viewing.

In the early 1990s, one of Jupiter's belts seemed to disappear overnight. Later, it reappeared. If this happens again, it may well be an amateur astronomer who spots it first.

If you don't see the Great Red Spot at first, you may be looking at it when the spot's in a pale condition, but more likely, it's just on the backside of Jupiter. You have to wait for Jupiter to turn enough for the Red Spot to appear. If you take telescopic views of the features of Jupiter at intervals of an hour or two during the night, you should see these features move across the disk of the planet as Jupiter rotates.

Jupiter also has rings, which are thin and made of rock particles, perhaps tiny fragments from numerous boulders in the rings. They are dark and not visible in amateur telescopes. In fact, they are hard to see in any telescope except the Hubble and those instruments that are carried right up to Jupiter on space probes. Microscopic particles from the rings are lifted up into a thick halo interior to the rings, which surrounds and perhaps merges into the upper atmosphere of the planet.

Jupiter turns so fast that rotation makes it bulge at the equator and flatten at the poles. With a clear look in steady air, you should be able to detect this *oblate* shape in your telescope.

Shooting for Galileo's moons

Whenever the atmosphere is steady and you have a good view through your telescope, you can see structure on Jupiter, and you're likely to see one or more of the planet's four large moons — Io, Europa, Ganymede, and Callisto.

Jupiter's four prominent moons are known as the Galilean moons, or Galilean satellites, named after Galileo, their discoverer. Each of the big four moons orbits almost exactly in the equatorial plane of Jupiter. So each moon is always right overhead somewhere on Jupiter's equator. Any telescope worthy of the name will show the Galilean moons, and many people can even see two or three of them through a good pair of binoculars. Io, the innermost of the Galilean moons, is hard to spot through binoculars because it's always very close to the bright planet. (Jupiter also has at least 12 smaller moons.)

You can't see enough detail on any of Jupiter's (or Saturn's) moons with your own telescope to figure out what their surfaces are like, but you will notice differences in their brightnesses and, with careful study, perhaps in their colors as well.

But if you take a look at the space probe images of the Galilean moons, you'll see that each moon is a little world unto itself, with composition and landscape that give them individual character.

- At 3,274 miles in diameter (5,268 kilometers), *Ganymede* is larger than 3,033-mile-wide Mercury (4,880 kilometers) and is the largest moon in the solar system. Ganymede's blotchy surface consists of light and dark terrains, perhaps ice and rock, respectively. The most noticeable marking is Valhalla, a huge ringed impact basin, about as large as the continental United States (judging the size by the outermost ring-ridge).

- *Io's* surface is peppered with more than 80 active volcanoes. It's the only place other than Earth where we have definite evidence of ongoing volcanism. Most likely, the volcanoes on Mars are long dead, and the evidence for active volcanism on Venus is very controversial — big volcanoes are discernible, but they are probably dead, too.

- *Europa* has a ridged terrain that looks like rafts of ice. The surface may be a frozen crust that tops off an ocean of water and slush, perhaps 90 miles (150 kilometers) deep. It is the only place in the solar system, outside of Earth, where we have strong evidence that liquid water is present. The existence of liquid water on Mars, beneath a layer of permafrost, is only a theory.

- *Callisto* has a dark surface, with many white craters. This surface is probably dirty ice, a mixture of ice and rock. Where asteroids, comets, and big meteoroids struck, the underlying clean ice has been exposed. Hence, the white craters.

Although you're not going to enjoy the kind of close-up and personal view that's afforded by sophisticated space equipment, you can expect to encounter a few aspects of these moons through your telescope lens (as I describe in the next section).

Following a moon shadow, and other view blockers

Io, Ganymede, Europa, and Callisto are always moving, changing their relative positions and appearing and disappearing as they revolve around Jupiter.

If you can't spot one of these moons, here are a few possible explanations:

✔ An *occultation* may be underway, which occurs when one of the moons passes behind the limb of Jupiter.

✔ The moon may be in *eclipse,* which occurs when the moon moves into Jupiter's shadow. Because the Earth is often well to the side of a straight line from the Sun to Jupiter, Jupiter's shadow can extend well to the side of Jupiter, as seen from the Earth. When a moon that's in plain view, off the limb of Jupiter, suddenly dims and disappears, it's gone into Jupiter's shadow.

✔ The moon may be *in transit* across the disk of Jupiter; at that time, the moon is particularly tough to see. That's because the moons are pale in color, making them hard to spot against the cloudy atmosphere of Jupiter. In fact, the moon in transit can be much more difficult to discern than its shadow.

You can also observe a *moon shadow,* which occurs when one of the moons is sunward of Jupiter and casts a shadow on the planet. The shadow is a black spot, much darker than any cloud feature, moving across the planet. The moon that casts the spot may be in transit at the time, meaning that as seen from Earth, it is viewed against the disk of Jupiter. But this is not always the case. When the Earth is well off the Sun-Jupiter line, a moon that is off the limb of Jupiter may be seen to cast a shadow on the planet.

Timing your moon gaze just right

Sky & Telescope magazine carries a monthly schedule of the occultations, eclipses, shadow events, and transits of the four Galilean moons. Both that publication and *Astronomy* also print a monthly chart showing the positions of the four moons with respect to the disk of Jupiter night by night. You can tell which moon is which by comparing what you see through the telescope with the chart. Remember these general rules as you watch or anticipate occultations, eclipses, shadows, and transits:

✔ All four Galilean moons orbit around Jupiter in the same direction. When they are on the near side, as seen from Earth, they move from east to west, and when they are on the far side, they are traveling from west to east.

✔ So a transiting moon is moving westward, and a moon that is about to be occulted or eclipsed is heading eastward. These are the east-west geographic directions in the sky of Earth.

Under excellent viewing conditions, observers with 6-inch or larger telescopes can even discern markings on Ganymede, the largest Galilean moon. (See Chapter 3 for information on telescopes.) But to see the details of the surface, you need an image from an interplanetary spacecraft that visited the Jupiter system.

The best images of Jupiter and its moons are from the Galileo and Voyager 1 and 2 space probes and from the Hubble Space Telescope. Galileo is the newest and most advanced Jupiter probe; you can find its images at galileo.ivv.nasa.gov/images.html. For Hubble images, examine the collection at the Space Telescope Science Institute, at oposite.stsci.edu/pubinfo/SolarSystemT.html#Jupiter. The Voyager images, plus some others, are at NASA's Planetary Photojournal Web site, photojournal.jpl.nasa.gov/. When you see the planets, just click on the picture of Jupiter.

Ring Up Your Friends to See Saturn!

Most people are familiar with the planet Saturn because of its striking collection of rings. For centuries, astronomers thought that Saturn was the only planet that has rings. Today, we know that rings encircle all four giant gas planets: Jupiter, Saturn, Uranus, and Neptune. But most of those rings are too dim to be seen through small telescopes or even large telescopes from the ground. The great exception is Saturn!

Com(et)ing within striking distance

On rare occasions, a comet strikes Jupiter, which causes a temporary dark blotch that may last for months. No one knew this until July 1994, when large chunks of the broken comet Shoemaker-Levy 9 struck Jupiter. But astronomers have gone back through old records of the markings on Jupiter and have found some suspicious features that may have been created in the same way.

The odds are poor that you will witness a comet striking Jupiter, but keep the possibility in mind. If you see any new dark blotch, make good notes.

The Arizona-based Canadian amateur astronomer David Levy achieved international fame after he helped discover the Jupiter-smashing comet Shoemaker-Levy 9. Thanks to his lucid accounts of this and other astronomical events, he now commands lucrative fees for appearances, articles, and books. His writing on celestial stars appears regularly in *Parade,* alongside other writers' personality profiles of the Hollywood variety. You, too, may enjoy that kind of notoriety — just keep a close watch on solar system traffic!

Saturn's rings are usually easy to see because they are large and composed of bright particles of ice — millions of little ice fragments, some larger ice balls, and maybe some that are the size of boulders. You can enjoy the rings in a small telescope, and also make out their shadow on the disk of Saturn (see Figure 8-3). Under excellent viewing conditions, the *Cassini division* — a gap in the rings that's named for the person who first reported it — may also be evident.

Figure 8-3:
Saturn and its rings.

Photo Courtesy of NASA

The 17th century astronomer Galileo Galilei, who discovered Saturn's rings, was mystified when they seemed to disappear briefly. Of course, what happened was that after many nights of viewing, he caught Saturn when its rings were edge-on to Earth (see the next section, "Cry 'tilt' if you don't see the rings," to see what a difference an edge makes). The rings aren't visible when their edges face Earth because they are extremely thin.

Measuring more than 124,000 miles (200,000 kilometers) across, Saturn's rings are only tens of yards (or meters) thick. Proportionately, the rings are comparable to "a sheet of tissue paper spread across a football field," as described by Professor Joseph Burns of Cornell University. But even though they're proportionately as thin as a huge facial tissue, you wouldn't want to blow your nose in them. Stuffing ice up your nostrils may chill you out more than sniffing glue, but it's definitely not recommended.

Cry "tilt" if you don't see the rings

Sometimes Saturn's rings are hard to discern in the same telescope that revealed them in splendor just a few months before. They can even seem to vanish in small telescopes, when we see them nearly edge-on.

The rings are very large, but very thin. They keep a fixed orientation, pointing face-on at one direction in space. There's a time each year when the rings are roughly face-on, as visible from Earth, and a time three months later when they are roughly edge-on, and this cycle repeats.

As Saturn goes around its own 30-year orbit, however, there are times — every 15 years — when the rings are precisely edge-on and seem to vanish in small (or sometimes even large) telescopes. On those occasions, with a powerful telescope, you may see the rings projected as a very thin dark line against the disk of Saturn. That last happened in 1996, so don't worry about missing a chance to see the rings until the next vanishing act in 2011.

Watch out for storms!

Saturn also has belts and zones, just like Jupiter, but they have less contrast and are much harder to see. Look for them during times of good atmospheric conditions, when you can use a higher power eyepiece to spot planetary details.

About once every 30 years, a big white cloud or "great white storm" appears in the Northern Hemisphere of Saturn. High-speed winds spread the cloud out until it forms a thick, bright band all the way around the planet. Months later, it's all gone. Sometimes amateur astronomers are the first to spot a new storm on Saturn. The last great white storm was in 1990, so you may have to wait a long while to see another. In the meantime, keep an eye out for smaller white cloud spots that spread partway around the planet. Saturn spins once every 10 hours, 39 minutes, and 22 seconds and is even more oblate — flattened at the poles — than Jupiter. The rings tend to mislead the eye a little, however, and seeing Saturn looking squashed can be tricky.

A moon of major proportions

Titan, Saturn's largest moon, is bigger than the planet Mercury. Its diameter is 3,200 miles (5,150 kilometers). Some of Jupiter's other large moons have very thin atmospheres, but Titan has a thick, hazy atmosphere, composed of nitrogen and trace gases such as methane. This atmosphere is hard to see through, but images from the 400-inch (10-meter) Keck Telescope confirm that there are pronounced dark and light blotches on Titan's surface.

According to the leading theory, the dark areas on Titan are lakes or oceans of liquid hydrocarbons, such as ethane. If we sailed on a sea like that on Earth, the cruise ship would be strictly "No Smoking." We don't need another "Titanic" disaster. With a good small telescope, you may be able to see two other moons, Rhea and Dione, when they are near their largest elongations from the planet. You can find a monthly chart of the locations of the moons with respect to the disk of Saturn in *Sky & Telescope*.

The latest and best picture of Saturn's moon, Titan, is at the Lawrence Livermore National Laboratory Web site (www.igpp.llnl.gov/titan/images.html). The most outstanding images of Saturn overall are from Voyager 1 and 2 and from Hubble. Look for Voyager pictures at NASA's Planetary Photojournal; just click on the picture of Saturn. The Hubble images are at oposite.stsci.edu/pubinfo/SolarSystemT.html#Saturn.

The Cassini space probe is en route to Saturn and Titan. Check its progress at the Cassini Web site at www.jpl.nasa.gov/cassini.

According to many observers, Saturn is the most beautiful planet. Its famous rings are easily visible in almost any telescope, and Saturn's giant moon, Titan, can be spotted as well. Although many astronomers find Saturn's rings to be the celestial sight that most impresses their nonastronomer friends, Titan is also a worthy attraction.

Moons on the move

Jupiter has 16 known moons and Saturn has 18, at least as of mid-1999. Each planet probably has quite a few more small ones, and astronomers keep finding them. Any number you find in a printed book may be obsolete by the time that you read it. Sometimes moons are announced, but not counted. The officials at the International Astronomical Union want to be sure that the discoveries are confirmed.

The moons come in two varieties: regular and other. The regular moons are all orbiting in the equatorial plane of their planet, and they all orbit in the same direction that the planet spins on its axis. This direction is called *prograde*. The regular moons almost certainly formed in place around Jupiter and Saturn, from an equatorial disk of protoplanetary and protomoon material. So Jupiter and Saturn, together with their many moons, are like miniature solar systems, centered on big planets rather than stars.

But some of the small moons are like Elsa, the lioness that was "born free." They orbit in the direction opposite to the way their planets turn. Those orbits are *retrograde*. And these orbits may also be tilted with respect to the equatorial planes of their planets. Those moons formed elsewhere in the solar system, perhaps as asteroids, and were then captured by Jupiter or Saturn.

Chapter 9

Far Out! Uranus, Neptune, and Pluto

In This Chapter

▶ Understanding the great worlds of water and rock — Uranus and Neptune

▶ Questioning the nature of Pluto

▶ Entering the Kuiper Belt

▶ Observing the outer solar system

Although Mars and Venus are closer to Earth and Jupiter and Saturn are bright, showy objects, observing the outer planets has its own mystique and rewards. This chapter introduces you to our solar system's three outer planets — Uranus, Neptune, and Pluto — and their moons. Plus, I offer some useful tips for viewing these far-out worlds.

The Nature of Uranus and Neptune

The following are the most important facts about Uranus (pronounced *yoo-RAN-us,* or more commonly *YOO-rin-us*) and Neptune:

✔ They are two similar-sized planets with similar chemical compositions, and they are smaller and denser than Jupiter and Saturn.

✔ Each planet is the center of a miniature system of moons and rings.

✔ Each planet has apparently suffered a major encounter with another body long ago.

The atmospheres of Uranus and Neptune, like those of Jupiter and Saturn, are mostly hydrogen and helium. But Uranus and Neptune are termed *icy planets* by astronomers, because their atmospheres surround massive cores of rock and water. In fact, the water is so deep inside the planets and under such high pressure that it is all hot liquid. But when these planets merged and coalesced from smaller bodies billions of years ago, the water that came into them was all frozen.

You can tell a bona fide planetary scientist from a lay person because the scientist calls the hot water inside Uranus and Neptune "ice," but the civilian innocently calls hot water "hot water." Scientists use technical jargon the way predatory animals use scent markings, to proclaim their exclusive territory.

Uranus has about 14.5 times the mass of Earth, and Neptune equals 17.2 Earths, but they are nearly the same size. The lighter Uranus is a bit larger, measuring 31,770 miles (51,118 kilometers) across the equator. Neptune's equatorial diameter is 30,784 miles (49,532 kilometers).

One day on Uranus is about 17 hours and 14 minutes; a day on Neptune is 16 hours and 7 minutes. So, like Jupiter and Saturn, these planets both rotate much faster than Earth.

The bull's-eye: Tilted Uranus and its rings and moons

The evidence that Uranus suffered a major collision or gravitational encounter is that the planet seems to have flipped on its side. Instead of the equator being in roughly the plane of Uranus's orbit around the Sun, it's at nearly right angles to that plane, so that, in terms of Earth's directions, Uranus's equator runs roughly north-south.

Sometimes the North Pole of Uranus points toward the Sun and Earth and sometimes (as now) the South Pole faces that way. One revolution around the Sun, Uranus's year, is almost 84 Earth years. For about one-quarter of that time, the North Pole faces roughly sunward; for about another quarter the South Pole faces roughly sunward; the rest of the time, the Sun shines down on the equator.

On Earth, the Sun is never high in the sky at the North Pole or the South Pole, but on Uranus it sometimes passes overhead at the poles.

As of late 1999, Uranus had 17 known moons and four other reported moons, not yet confirmed. Uranus also has a set of dark rings. The moons and rings of Uranus orbit in the plane of its equator, just as the Galilean moons orbit in the equatorial plane of Jupiter (see Chapter 8). So the rings and the orbits of the Uranus moons are at nearly right angles to the plane of Uranus's orbit around the Sun.

You can think of the Uranus system (the planet and its satellites) as a big bull's-eye, which sometimes faces Earth and sometimes not. Something hit that bull's-eye long ago and tilted it over from its natural position.

Neptune and its backward moon

Neptune is not tilted from the natural order; its equator is in the plane of its orbit, or nearly so. Neptune has eight known moons, as of mid-1999. But its largest moon, Triton (which is larger than Pluto), with a diameter of 1,684 miles (2,710 kilometers), is in retrograde orbit. Seen from north and above, Neptune, like all the planets in our solar system, revolves counterclockwise around the Sun. And most moons revolve counterclockwise around their planets. But Triton, which looks a lot like a cantaloupe in photos from Voyager 2, goes against the grain, travelling clockwise around Neptune. After mulling (or meloning) this over, scientists have concluded that Triton came too close to Neptune long ago and was captured, becoming a moon when it might otherwise have been a planet a lot like Pluto.

Triton *is* made of real ice and rock. So it's much more like Pluto than like Uranus and Neptune. Triton's surface is shaped by *cryovolcanism,* meaning eruptions and flows of cold icy substances, rather than hot, molten rock. Water ice, dry ice, frozen methane, frozen carbon monoxide, and even frozen nitrogen are all present on Triton. Triton doesn't have many impact craters, probably because they get sloshed full after a while.

Environmental groups say that national parks such as Yellowstone are endangered by excessive tourism. So consider a trip to Triton instead. Its landscape is just as bizarre, and maybe as beautiful, as Yellowstone's. But if you are heading for Triton, take your booties and duck the geysers! There are cold surges instead of hot springs, and Triton's geysers spew long plumes of frigid, dirty vapor rather than torrid jets of steam. But there's plenty of parking on Triton and no bears to invade your picnic supplies. Just bring a space suit and those extra-warm booties.

Pluto Is No Comic Character

Pluto is the smallest planet and the most distant one (see Figure 9-1). It comes inside the orbit of Neptune every 248 years for a few decades at a time, but the last such inside move ended in early 1999. It won't happen again in the lifetime of anyone now on Earth, unless medical research makes major strides between now and the 23rd century.

Pluto takes 6 days 9 hours 17 minutes to turn once on its axis, and its moon Charon goes once around the planet in exactly the same amount of time. So the same hemispheres of Pluto and Charon always face each other. In the Earth-Moon system, one hemisphere of the Moon always faces Earth, but not vice versa. Someone on the near side of the Moon can see the whole Earth over the course of an Earth day, but a person on Charon could never see more than half of Pluto.

Figure 9-1:
The odd
little planet,
Pluto.

Photo Courtesy of NASA

Pluto is 1,430 miles (2,300 kilometers) in diameter, making it the smallest planet. It's also smaller than the four Galilean moons of Jupiter and the moons Titan of Saturn and Triton of Neptune. In fact, Pluto is less than twice as big as 780-mile- (1,250-kilometer-) wide Charon. So Pluto and Charon are often called a double planet.

Pluto and Charon are both icy, rocky worlds, and unlike Uranus and Neptune, this ice is real, not molten.

With a surface temperature of −387° F (−233° C), it's hardly surprising that almost everything freezes on Pluto. Water ice, methane ice, nitrogen ice, ammonia ice, and even frozen carbon monoxide are present on the surface of Pluto. It exhausts me to think of it! Some, but not all, of these substances have been detected on Charon.

Pluto is not quite as cold as it may seem. Astronomers suspect that it has some "tropical oases" where the temperature gets all the way up to −351° F (−213° C).

Pluto is so far away that scientists have little idea about its geography. Its elongated elliptical orbit takes it within about 29.7 A.U. or 2.8 billion miles (4.4 billion kilometers) of the Sun and as far out as 49.5 A.U. or 4.6 billion miles (7.4 billion kilometers).

Images from the Hubble Space Telescope (see them at `oposite.stsci.edu/`
`pubinfo/SolarSystemT.html#Pluto`, along with an animation of the turning globe of Pluto) show lighter and darker regions, which may correspond to
places where there is fresh ice and old ice, respectively, but that's about all.
No space probe has ever visited Pluto, and although NASA has plans for a
possible mission to Pluto (see the Web site of the Jet Propulsion Laboratory,
`www.jpl.nasa.gov/ice_fire//pkexprss.htm`), the official go-ahead has
not been given yet.

Pluto, like Uranus, is tilted on its side, with its rotational axis (the line
through the North and South Poles) roughly perpendicular to its orbit. So it,
like Uranus, has probably suffered a major collision. In fact, some
astronomers believe that Charon is a chip off Pluto's block, created by an
impact on the planet — much as the Moon is believed to have formed from a
great impact on Earth (see Chapter 5).

Is Pluto a Planet?

Every so often someone casts aspersions on Pluto and suggests that it
shouldn't count as a planet. Most recently, in 1999, there was an attempt to
designate it as asteroid No. 10,000. But astronomers and just plain folk rallied
round the little cold body and defeated this plan. They argued that it's round
like a planet (most asteroids, other than the largest ones, are irregularly
shaped), it has a large moon, and it has been considered a planet from the time
it was discovered by American observer Clyde Tombaugh in 1930. Even if
astronomers change the definition of a planet, Pluto should be grandfathered in.

Pluto is out in the Kuiper Belt, a region beyond Neptune where small icy
bodies abound. It's estimated that about 100,000 Kuiper Belt Objects (KBOs)
larger than 60 miles (100 kilometers) in diameter are located between
Neptune's orbit and a distance of 50 A.U. from the Sun. They are beyond the
reach of backyard telescopes, unless your backyard is on Neptune or one of
its moons. The first KBO was discovered in 1992, and since then more than
150 have been found. The astronomers who thought Pluto should be down-
graded from a planet considered it to be the largest KBO. But it can be a big
KBO and a planet, too.

What are Plutinos?

The region of the Kuiper Belt has not been thoroughly surveyed by
astronomers, and experts calculate that among the thousands of KBOs that
are yet to be discovered and studied, one or two may be as big as Pluto. They
might be dimmer than Pluto because they have darker surfaces and/or are
further from the Sun. The discovery of one of these big ones is sure to cause
a controversy over whether to call it a planet.

Among the over 150 known KBOs are some that share three properties with Pluto:

- ✔ They have highly elliptical orbits.

- ✔ Their orbital planes are tilted by a significant angle with respect to the plane of Earth's orbit.

- ✔ They make two complete orbits around the Sun in approximately the same time that Neptune takes to make three orbits (496 years for Pluto's two orbits and 491 years for Neptune's three). This effect is called a *resonance*, and it works to keep Pluto and Neptune from ever colliding or even coming close to each other, although their orbits cross. So Pluto is safe from disturbance by the powerful gravity of much larger Neptune, and so are the KBOs that share these three properties.

The KBOs that share these properties are called *Plutinos*, meaning little Plutos.

There are probably other kinds of objects beyond Neptune and Pluto that astronomers have not yet discovered, but the objects can't be very massive, or their gravitational effects on known objects would have been detected by now. The only large planets beyond Neptune and Pluto are the planets of other stars. Chapter 15 covers those planets.

You can find out more about KBOs on the Nine Planets Web site at `seds.lpl.arizona.edu/nineplanets/nineplanets/kboc.html`.

The Outer Planets: Viewing Challenges

With experience, you can locate the large outer planets Uranus and Neptune, but tiny Pluto may be beyond your visual reach. The first time you look for any of these planets, you will do best with the aid of a more experienced amateur astronomer.

Sighting Uranus

Uranus was discovered with a telescope, but it sometimes is bright enough to be barely visible to the eye under excellent viewing conditions. In your telescope, you can distinguish Uranus from a star thanks to

- ✔ Its small disk, a few seconds of arc in diameter (I define this unit in Chapter 6)

- ✔ Its slow motion across the background of faint stars

The disk of Uranus has a pale green tint; you can make out the disk with a high-power eyepiece when viewing conditions are good. You can detect the motion of Uranus by making a sketch of its relative position among the stars in the field of view. For this purpose, use a low-power eyepiece so that the field of view is larger, and more stars are visible. Then look again in a few hours or on the following night, and sketch again.

As of late 1999, there were 17 known moons of Uranus and four others have been reported, but not yet confirmed. Although a few of the biggest moons can be glimpsed with large amateur telescopes, they are all objects for study with powerful observatory telescopes. Uranus also has a set of dark rings, detectable with the Hubble Space Telescope and in images made in infrared light with large telescopes on Earth.

You can see the Hubble Space Telescope images of these bodies at oposite.stsci.edu/pubinfo/SolarSystemT.html#Uranus. You can browse through images of Uranus and its moons from the Voyager 2 space probe at photojournal.jpl.nasa.gov, the Planetary Photojournal Web site. Just click on Uranus.

Distinguishing Neptune from a star

Neptune is fainter than Uranus, but gets as bright as 8th magnitude. If Uranus challenges your observing skills, buckle up for Neptune!

Neptune is about the same actual size as Uranus, but is much further away, so when viewed through a telescope its apparent disk is smaller. You may need a large amateur telescope to discern it from a star. If you are really good at perceiving pale hues in dim objects seen through a telescope, you will notice that Neptune has a blue tint.

Because Neptune is further from the Sun than Uranus, it orbits the Sun at a slower speed. The slower speed, combined with a greater distance from Earth, means that the angular rate of speed across the sky — in arc seconds per day — is *usually* less for Neptune than Uranus. So you may have to wait another night or two to be sure that you have seen it move across the pattern of background stars.

I wrote "usually" because both Uranus and Neptune, like all the planets beyond Earth's orbit, show retrograde motion at times, just as Mars does (see Chapter 6). So these planets seem to slow down and reverse direction now and then. If you happen to catch Uranus when it is changing direction in the sky, its apparent motion will be much slower than usual, and by comparison, Neptune may be going full tilt at that time.

The annual *Observer's Handbook* of the Royal Astronomical Society of Canada (www.rasc.ca) always carries good maps of the changing positions of Uranus and Neptune during the year. Consult them to know where these planets are and when they will be reversing course. You can find similar maps from time to time in the astronomy magazines. (Check out *Astronomy* Magazine at www.astronomy.com and *Sky & Telescope* at www.skypub.com/sights/sights.shtml.)

Neptune has eight known moons, as of mid-1999. The largest is Triton. Once you have mastered locating Neptune, look for Triton with a telescope of 6 inches or more in diameter on a clear dark night. It's in a large orbit, ranging about 8 to 17 arc sec from Neptune (about four to eight Neptune diameters), so Triton might be mistaken for a star. But by sketching Neptune and the faint "stars" around it on successive nights, you can deduce which "star" is moving with Neptune across the starry background as it also moves around Neptune. It takes Triton almost 6 days to make one full orbit around the planet.

You can browse through images of Neptune and its moons from the Voyager 2 space probe at photojournal.jpl.nasa.gov, the Planetary Photojournal Web site. Just click on Neptune. And you can see the Hubble Space Telescope images of these bodies at oposite.stsci.edu/pubinfo/SolarSystemT.html#Neptune.

Striving to see Pluto

Pluto is a much tougher challenge than any other planet in the solar system. It's far away and very small. Typically, Pluto is just 14th magnitude. And it's moving away from the Sun and Earth and will be doing so for many years, as it traverses its 248-year orbit.

Skilled amateurs say that they have seen Pluto with 6-inch telescopes, but use the largest telescope you can find or borrow. I recommend that you use at least an 8-inch telescope. A finder chart for Pluto is printed each year in the *Observer's Handbook* of the Royal Astronomical Society of Canada (www.rasc.ca). And look for articles and observing advice in *Sky & Telescope* and *Astronomy* magazines.

Pluto's moon, Charon, is very close to Pluto, revolving around it in just 6 days 9 hours 17 minutes, and cannot be distinguished in any but the most powerful observatory telescopes.

Part III
Old Sol and
Other Stars

The 5th Wave   By Rich Tennant

"I'm glad you were able to see a white dwarf and a red giant last night. I just hope you remembered not to stare."

In this part . . .

This part introduces you to the stars. No, not those wealthy people in Hollywood — I'm talking about Earth's Sun and all the other stars in the Milky Way galaxy and beyond. You can find out about the types of stars and their life cycles, from birth to death. When Madonna and Brad Pitt are long forgotten, Alpha Centauri will still be shining.

I also provide a chapter on black holes and quasars, and I simplify these subjects so you won't get a migraine trying to understand them. However, the information about space and time distortions may twist your thoughts a bit.

Chapter 10

The Sun: Star of the Earth

In This Chapter

▶ The size, shape, and cyclic changes of the Sun

▶ Solar effects you may encounter

▶ Tips for viewing the Sun safely

▶ Eclipses and when to expect them

*A*lthough many people are attracted to astronomy by the beauty of a moon-lit night and a starry sky, you need nothing more than a sunny day to experience the full impact of an astronomical object firsthand. The Sun is the nearest star to Earth and, in fact, provides the energy that makes life possible.

The Sun is so commonplace in day-to-day life that people take it for granted. You may worry about getting sun burned, and the effects of ultraviolet rays on your skin, but you probably seldom think of the Sun as a primary source of absolutely stellar information about the nature of the universe. In fact, however, the Sun is one of the most interesting and satisfying astronomical objects to study, whether with backyard telescopes or advanced observatories and instruments in space. The Sun changes hour-by-hour and day-by-day. And you can show it off to the kids without keeping them up past their bedtime!

But don't even think about looking at the Sun, let alone showing it to a child or anyone else, without taking the proper precautions that I explain in this chapter. You don't want your view of the Sun to cost you your sight. Safety in viewing is the prime consideration; once you have that down pat, you can follow the Sun not only on a daily basis, but also over the 11-year sunspot cycle I describe later in this chapter.

This chapter introduces you to the science of the Sun, the Sun's effects on Earth and on industry, and to safe solar observing. This chapter will help you look at the Sun in a new way — safely, and with awe.

Don't Make Galileo's Blinding Mistake: Protect Your Sight from the Sun

The 17th-century Italian astronomer Galileo Galilei made the first great telescopic discovery about the Sun. By watching the daily movements of sunspots across the solar surface, he found that the Sun turns. But he also made a terrible mistake. He looked at the Sun through his telescope. This error caused him severe eye damage.

A telescope or binoculars collects more light than the human eye, and focuses it on a small spot on your retina. That intensity is fine if you are viewing a dim star or a planet, but it's a prescription for eye damage or blindness if you look at the Sun with your instrument.

Ever see a "burning glass," a magnifying lens used to focus the rays of the Sun on a piece of paper to set it on fire? Now you get the idea.

Taking even the briefest peek at the Sun through a telescope, binoculars, or any other optical instrument is very dangerous unless the device is equipped with a solar filter made by a reputable manufacturer specifically for viewing the Sun.

I talk about filters and other safe-viewing techniques later in this chapter. But first I want to introduce you to the Sun itself, and to the nature of the fascinating sights you may observe.

Surveying the Sunscape

The Sun is a star, a hot ball of gas shining under its own power with energy from *nuclear fusion,* the process by which the nuclei of simple elements are combined into more complex ones. This energy powers not only the Sun itself, but much of the activity in the system of planets and planetary detritus that surrounds the Sun — the solar system of which Earth is a part (see Figure 10-1).

The Sun produces energy at an enormous rate, equivalent to the explosion of 92 billion one-megaton nuclear bombs every second. That energy comes from the consumption of fuel. If the Sun were made of burning coal, it would burn up every last lump of itself in just 4,600 years. But fossil evidence on Earth shows that the Sun has been shining for more than 3 billion years, and astronomers are certain that it's been going for longer than that. The estimated age of the Sun is 4.6 billion years, and it's still burning strong.

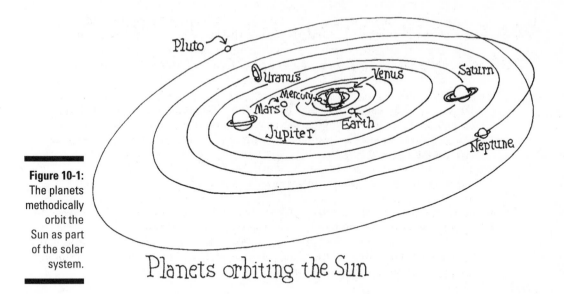

Planets orbiting the Sun

Figure 10-1:
The planets methodically orbit the Sun as part of the solar system.

Only nuclear fusion could produce the Sun's huge total energy release — its *luminosity* — and keep it going for billions of years. Near the center of the Sun, the enormous pressure and the central temperature of almost 16 million degrees Celsius (29 million degrees Fahrenheit) cause hydrogen atoms to fuse into helium, a process that releases the great torrent of energy that drives the Sun.

About 700 million tons of hydrogen turn into helium every second near the center of the Sun, with 5 million tons vanishing, turned into pure energy.

If we could generate energy that way on Earth, all our problems with fossil fuel, including air pollution and the consumption of nonrenewable resources, would be solved. But despite decades of research, scientists still can't do what the Sun does naturally. Clearly, the Sun deserves further study.

The Sun's size and shape: What keeps all those hot gases balled up together?

When I taught Astronomy 101, I'd always pose the question, "Why is the Sun the size that it is?" Hundreds of mouths would drop open, dozens of pairs of eyes would wander the room, but hardly ever would anyone have a clue. It doesn't even seem like a logical question. Everything has a size doesn't it? So what?

But if the Sun is made of 100-percent hot gas, what's keeping it together? Why doesn't it all blow away, like a smoke ring from the old cigarette sign that loomed over Times Square? The answer, my friend, is that gravity keeps the Sun from blowing in the wind. Gravity is the force, which I describe in Chapter 1, that affects everything in the universe. The Sun is so massive — 330,000 times the mass of planet Earth — that its powerful gravity can hold all that hot gas together.

Well, you might ask, if the Sun's gravity is pulling it all together, why isn't it squeezed down into a much smaller ball? The answer to that one is the same thing that sells many a used car: high pressure. The hotter the gas, and the more it's squeezed together by gravity or any other force, the higher its pressure. And gas pressure tends to inflate the Sun, just like it inflates an automobile tire (in which the gas is just air).

Gravity pulls in; pressure pushes out. At a certain diameter, the two opposite effects are equal and in balance, maintaining a uniform size. So the Sun is the size that it is. Its diameter is 864,500 miles (1,391,000 kilometers) or about 109 times the diameter of Earth. You could fit 1,300,000 Earths inside the Sun, but I don't know where you'd get them.

The Sun is round for much the same reason: Gravity pulls equally in all directions toward the center, while pressure pushes out equally in all directions. If the Sun were turning rapidly, it would bulge a little at the equator and be flattened slightly at the poles, from what people often call centrifugal force. But the Sun turns at a very slow rate, only once every 25 days at the equator (and slower near the poles), so any midriff bulge is very small.

The Sun's regions: Caught between the core and the corona

The Sun has two main regions on the inside, and three on the outside (see Figure 10-2). The inside of the Sun is called the stellar interior. At its center is the *core*. In the heart of the core, all the Sun's energy is generated by nuclear fusion. The energy is released in the form of gamma rays, a very energetic type of light. The gamma rays bounce around, off one atom and another, back and forth, but on average moving upward and outward. The further out in the solar interior, the cooler the temperature gets.

At a distance of about 307,000 miles or 494,000 kilometers (about 71 percent of the way from the center to the surface), the core gives way to the next major region, the *convection zone*. Here, huge streams of gas carry much of the upward-going energy. The hot gas streams rise up, bringing heat along; they cool with altitude and fall down again. It's the same process that brings heat from the bottom of a kettle of boiling water up to the surface and forms clouds in Earth's atmosphere. Solar physicists believe that the magnetic fields of the Sun, which cause sunspots and explosions of many kinds in the Sun's upper atmospheric regions, are generated near the bottom of the convection zone.

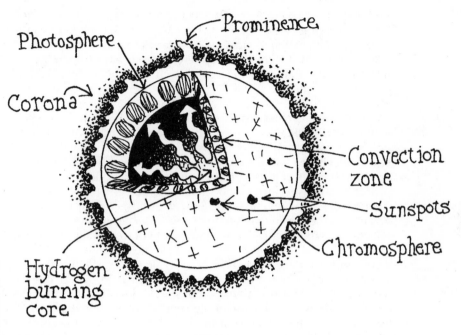

Photosphere

Prominence

Corona

Convection zone

Sunspots

Chromosphere

Hydrogen burning core

Figure 10-2:
The Sun is a hotbed of activity as it powers its piece of the universe.

Layers of the Sun

When you get to the heart of the matter, there are divisions within the Sun's core. The inner or *energy-generating core* extends to 108,000 miles from the center. Above that, the rest of the core is called the *radiative zone.*

The temperature of the convection zone drops from about 2.2-million° Celsius (4-million° Fahrenheit) at its bottom until it reaches the temperature of the next region of the Sun, which is the visible surface called the *photosphere* (which means "sphere of light"). It's the layer of gas, with a temperature of about 5,500° Celsius (9,900° Fahrenheit) that produces all the visible light of the Sun, except what we can see at the time of a total eclipse of the Sun (or with special instruments). Dark spots seen on the photosphere are called *sunspots,* and they are the most easily observable solar features.

If you look at the bright disk of the Sun — do so only in accord with the safety instructions in the section, "Project or Filter to Protect: Safe Techniques for Solar Viewing," later in this chapter — you are seeing a portion of the photosphere.

Above the photosphere, the successive regions of the Sun are hotter, not cooler than those below. This fact is one of the Sun's great mysteries, which astronomers have been puzzling over for decades. The *chromosphere* or *color sphere,* is just above the photosphere. It's only about 600 miles (1,000 kilometers) thick, but the temperature of the chromosphere reaches 10,000° Celsius (18,000° Fahrenheit).

You can view the chromosphere at the edge of the Sun if you use one of those expensive H-alpha filters I mention in the sidebar "Seeing more when price is no object" later in this chapter, or you can see it on images taken with professional telescopes and displayed on the NASA and NOAA Web sites (see the section "Looking at solar pictures on the World Wide Web"), and on various professional observatory Web sites. And you can see the chromosphere at the time of a total eclipse of the Sun, as discussed later in this chapter. During an eclipse, the chromosphere may appear as a thin red band all around the edge of the Moon, which is blocking the light from the photosphere.

Above the chromosphere is the *corona,* a gas so rarified and electrified that its shape is determined by the Sun's magnetic field. Where lines of magnetic force stretch and open outward into space, the coronal gas is thin and barely visible. It can readily escape in the form of solar wind. Where lines of magnetic force reach up in the corona and then turn back down to the surface, they confine the coronal gas. It's thicker and brighter there. The corona is a sizzling 1-million° Celsius (1.8 million° Fahrenheit) and in places it's even hotter.

The transition from the chromosphere to the hundred-times-hotter corona occurs in a very thin boundary layer called the *transition region.* It's not evident in views of the Sun.

The solar wind: Playing with magnets

Solar wind is an electrified gas or plasma that moves out through the solar system at about 470 kilometers per second (1-million miles per hour) as it passes Earth's orbit.

The solar wind comes in streams, fits, and puffs and constantly disturbs and replenishes Earth's magnetosphere. (The *magnetosphere* is a huge region around Earth in which electrons, protons, and other electrically charged particles bounce back and forth from high northern latitudes to high southern latitudes, trapped in Earth's magnetic field.) As I describe in Chapter 5, it was first known as the Van Allen Belts, after James Van Allen of the University of Iowa, who discovered it with the Unites States' first satellite, Explorer 1.

Thanks to the variable nature of the solar wind, and to traveling solar storms that pervade it after eruptions on the Sun, the magnetosphere is constantly disturbed, becoming compressed in size and swelling out again; its changes produce geomagnetic storms that disturb the environment on and above Earth.

Solar activity and solar cycles: How's the weather out there?

All kinds of disturbances take place on the Sun from moment to moment and from one day to the next, including many that occur in the vicinities of sunspot groups, which I discuss later in this chapter. Some of this *solar activity* actually affects Earth.

Solar eruptions — mostly invisible with amateur equipment, but marvelously revealed by satellite telescopes — spray billion-ton blobs of solar plasma, permeated with magnetic fields, out into the solar system, where some of them collide with Earth's protective magnetic umbrella, the magnetosphere. This activity can cause displays of the Northern Lights (aurora borealis) and Southern Lights (aurora australis) and geomagnetic storms. The geomagnetic storms can shut down power company utility grids (causing blackouts), blow out electronics circuits on oil and gas pipelines, interfere with radio communications, and damage expensive satellites. Some people even claim they can *hear* aurorae.

The solar disturbances and their effects on the magnetosphere are called *space weather.* You can see the latest official U.S. government space weather report and forecast at a Web site of the Space Environment Center, a unit of the National Oceanographic and Atmospheric Administration (`www.sec.noaa.gov/today.html`). All forms of solar activity, including the 11-year sunspot cycle and some even longer cycles, seem to involve magnetism. Deep inside the Sun, a natural dynamo is generating new magnetic fields all the time. The magnetic fields rise to the surface of the Sun and on up to higher layers in the solar atmosphere, where they twist around and cause all kinds of trouble.

Astronomers measure magnetic fields on the Sun by their effects on solar radiation, using instruments called *magnetographs.* You can see images taken with these devices on many of the professional solar observatory Web sites. (See the section "Looking at solar pictures on the World Wide Web) These magnetic field observations show that sunspots are areas of concentrated magnetic field, and that sunspot groups have north and south magnetic poles. On the other hand, the overall magnetic field of the Sun is pretty weak.

Many of the rapidly changing features on the Sun and probably all explosions and eruptions seem to be related to solar magnetism. Where there are changing magnetic fields, there are electrical currents, and when two magnetic fields bump into each other, a short circuit — called a *magnetic reconnection* — can suddenly release huge amounts of energy.

Coronal mass ejections: The mother of solar flares

What you are about to read is contrary to what is in most of the textbooks, except some published very recently. For decades astronomers believed that the primary explosions on the Sun were *solar flares*. We thought that solar flares occur in the chromosphere, and that they are what set things off.

You can see solar flares in many of the images on the professional astronomy Web sites. As the number of sunspots increases over an 11-year sunspot cycle, so does the number of flares.

Now astronomers know that they were just like the blind man who feels the elephant's tail and thinks he knows all about the elephant when he is only touching one of the beast's less significant parts. Observations of the Sun from space reveal that the primary engines of solar outburst are the *coronal mass ejections,* huge eruptions that occur high in the corona, the thinnest, outermost solar region. Often, a coronal mass ejection triggers a solar flare beneath it in the low corona and chromosphere.

Scientists didn't know about coronal mass ejections for many years because no one could see them. Astronomers could only get a good view of the corona at rare intervals during the brief duration of a total eclipse of the Sun. All they could see were the solar flares, so scientists attributed more importance to them than they deserved.

Some of the prominences that you can see on the edge of the Sun with an H-alpha filter occasionally erupt. These eruptive prominences may also be phases of coronal mass ejections.

When the satellite images show a coronal mass ejection that is not going off, say to the east or to the west from the Sun, but that forms a huge expanding ring or *halo event* around the Sun, that's bad news. That halo event means that the coronal mass ejection is heading right at Earth.

If you see a halo event in one of the satellite images, check the NOAA Space Environment Center Web site (www.sec.noaa.gov/today.html), because NOAA may be forecasting some pretty fierce space weather.

Cycles within cycles: Can the Sun change its spots?

Sunspots are areas visible as dark spots on the Sun's photosphere (see Figure 10-3) where the magnetic field is strong. Cooler than the surrounding atmosphere, sunspots often appear in groups.

The number of sunspots on the Sun varies dramatically over a repeating cycle that lasts about 11 years, the famous *sunspot cycle*. Before Americans had Richard Nixon or El Niño to kick around, everything from bad weather to a decline in the stock market was blamed on sunspots. Usually, 11 years pass between successive peaks (when more spots occur) of the sunspot cycle, but the period of time can vary. Further, the number of spots at the peak can vary widely from one cycle to the next. No one knows why.

As a sunspot group moves across the solar disk due to the rotation of the Sun, the biggest spot on the forward side (the one that leads the way across the disk) is called the *leading spot*. The biggest spot on the opposite end of the group is the *following spot*.

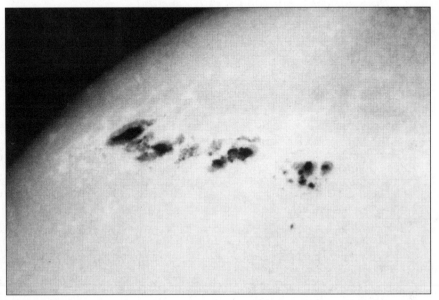

Figure 10-3: Sunspots.

Photo © Jerry Lodriguss

Magnetograph observations show definite patterns in most sunspot groups. During one 11-year sunspot cycle, all the leading spots in the Northern Hemisphere of the Sun will have north magnetic polarity while the following spots will have south magnetic polarity. At the same time, in the Southern Hemisphere, the leading spots will have south polarity and the following spots will have north polarity.

Here's how these polarities are defined: The compass needle that points north on Earth is called a north-seeking compass. A north magnetic polarity on the Sun is one that a north-seeking compass would point to. A south magnetic polarity on the Sun is one that a north-seeking compass would point away from.

Just when you think that you have all that straight, guess what? A new 11-year cycle begins, and the polarities are all reversed. In the Northern Hemisphere now, the leading spots have south polarity and the following spots have north polarity. In the Southern Hemisphere, the magnetic polarities have reversed, too. If you were a compass, you wouldn't know if you were coming or going.

To encompass all this information, astronomers have defined the Sun's *magnetic cycle*. It's about 22 years long, and contains two sunspot cycles. Every 22 years, the whole pattern of changing magnetic fields on the Sun repeats itself — more or less.

The solar constant: Why isn't it constant?

The total amount of energy produced by the Sun is called the *solar luminosity*. Of greater interest to us on Earth is the amount of solar energy that Earth receives, or the *solar constant.* Defined as the amount of energy falling per second on 1 square centimeter of area facing the Sun at the average distance of Earth, the solar constant amounts to 1,368 watts per square meter (127 watts per square foot).

Measurements made by solar and weather satellites sent up by NASA in the 1980s revealed very small changes in the solar constant as the Sun turns. You may think that less energy would be received when those dark sunspots are marking the solar disk than when they are not. But that's not the case; it's the opposite: more sunspots, more energy received from the Sun. That's another mystery for astronomers to solve.

According to astrophysical theory, the Sun was slightly brighter when it was very young than it has been for the last several billion years, and it will certainly cast more energy on Earth ages from now when it becomes a red giant star.

So "solar constant" is wishful thinking, although from day to day and with amateur equipment, it's pretty darn accurate.

The solar neutrino mystery: Why are some missing?

The nuclear fusion at the heart of the Sun does more than change hydrogen into helium and release energy in gamma rays to heat the whole Sun. It also releases enormous numbers of neutrinos, electrically neutral subatomic particles that have no mass (or almost no mass), travel at the speed of light (or nearly, depending on whether they actually do have mass), and can pass through almost anything.

A neutrino is like a hot knife in butter. It cuts right through.

In fact, neutrinos can fly right out from the center of the Sun and into space. Those that head Earthward fly right through Earth and out the other side. A few of those solar neutrinos are counted in huge underground laboratories known as neutrino observatories, located mostly in deep mines and tunnels under mountains, but a new one called AMANDA has been built into the thick Antarctic ice.

Counting neutrinos isn't easy, but reports from the neutrino observatories indicate a deficiency in solar neutrinos: The number of neutrinos coming through Earth is significantly fewer than the number that are expected based on the rate at which the Sun generates energy.

This solar neutrino deficiency is the least of our problems on Earth. It pales into insignificance beside food shortages in Africa, the depletion of the forests, the extinction of valuable species, and the consumption of irreplaceable fossil fuel reserves.

But the loss nags at scientists, prompting them to make new theories of particle physics and to check on theoretical models of the solar interior. As the saying goes, for want of a nail, the shoe was lost — and the deficiency of a bunch of neutrinos could be telling scientists something that will overturn the current paradigm of some area of physics or astronomy.

You'll have to wait and see how the case of the missing neutrinos pans out, but you can bet that astronomers will keep studying the Sun as long as mysteries such as the solar neutrino deficiency continue to confound them.

The Sun's lifespan: Will it ever die?

Some day, the Sun must run out of fuel, so some day it will die. All good things must come to an end.

Without the energy and warmth of the Sun, life on Earth would be impossible: the oceans would freeze, and so would the air. But what will actually happen is that the Sun will swell up and take the form of a red giant star. It will look enormous, and it will fry the oceans. So the oceans will actually evaporate before they have a chance to freeze.

Read the preceding paragraph carefully: I didn't say that the oceans will freeze; I said that they would freeze without the energy of the Sun. In fact, the energy received will increase so much before the Sun dies out that we'll die of the heat (if people still exist), not of the cold. Talk about global warming!

The giant red future Sun will puff off its outer layers, forming a beautiful expanding nebula. It will be the kind of shining gas cloud that astronomers call a planetary nebula. But no one will be here to admire it. So to appreciate what we will miss, take a good look at some of the planetary nebulas created by other suns; I describe them in Chapters 11 and 12.

The nebula will gradually clear away and all that will be left at its center is a tiny cinder of the Sun, a hot little object called a white dwarf star. It won't be much larger than Earth. Although it will be pretty hot, it will be so small that it casts little energy on Earth. So whatever's left on Earth will freeze. And the white dwarf will just be shining like the embers in a dying campfire. It will slowly fade away.

Fortunately, we have about five billion years to go before that prospect looms near. Future generations can worry about this, along with the national debt and how to acquire rare first editions of *Astronomy For Dummies*.

Project or Filter to Protect: Safe Techniques for Solar Viewing

Galileo was no fool. After he learned the hard way not to look at the Sun through a telescope, he invented the *projection technique,* using a simple telescope to cast an image of the Sun on a screen in the manner of a slide projector. This technique is safe only when used properly with simple telescopes, such as those sold under the description *Newtonian reflector,* or *refractor.*

As I explain in Chapter 3, a Newtonian reflector uses only mirrors, aside from the eyepiece, and its eyepiece is near the top of the telescope tube, protruding from it at right angles. A refractor works with lenses and doesn't contain a mirror.

Don't use the projection technique with telescopes that incorporate both lenses and mirrors in addition to eyepieces. In other words, don't use the projection technique with the Schmidt-Cassegrain and Maksutov-Cassegrain telescope models — including the highly regarded Meade ETX-90/EC telescope — which use both mirrors and lenses (I describe all these telescopes in Chapter 3). The hot, focused solar image may damage apparatus inside the sealed telescope tube, and could then pose a danger.

Viewing the Sun by using projection

Here's how to safely view the Sun by projection:

1. **Mount a Newtonian reflector or refractor telescope on a tripod.**

2. **Install your lowest-power eyepiece in the telescope.**

3. **Point the telescope in the rough direction of the Sun *without* sighting through or along the telescope; keep yourself and all other persons away from the eyepiece and not in line with it.**

4. **Find the *shadow* of the telescope tube on the ground.**

5. **Move the telescope up and down and back and forth while watching the shadow, to make the shadow *as small as possible*.**

   The best way to do this step is for you or an assistant to hold a piece of cardboard beneath the telescope, perpendicular to the long dimension of the telescope, so that the telescope tube shadow falls on the cardboard. Move the telescope so that the tube shadow is as close to a solid dark circular shape as possible.

6. **Hold the cardboard at the eyepiece; the Sun will be in the field of view and its image appears on the cardboard.**

   If the Sun's image is not in view, the bright glare of the Sun will be visible on one side of the cardboard; in that case, move the telescope to move that glare, thus moving the Sun into view on the cardboard.

Figure 10-4 presents a diagram of this technique. The easiest and safest way to learn this technique is to consult an experienced observer from your local astronomy club.

Even though you avoid looking through the telescope, you have to beware of some other hazards of the projection method. I once saw a tough guy at a school in Brooklyn project the solar image with a 7-inch telescope. He didn't get his face near the eyepiece, but at one point, he moved his arm through the projected beam, very close to the eyepiece where the solar image is very small. It burnt a little smoking hole in his black leather jacket.

To avoid injury, don't look through the eyepiece at the Sun and don't let any part of your body, anyone else's body, or any property get in the projected beam of sunlight.

Now you can look for sunspots across the *solar disk,* the scientific term for the visible surface of the Sun that is facing toward Earth. If you (pardon the expression) spot some spots, look again tomorrow and the next day, and you will see them seem to move across the solar disk. In reality, although they may move a bit on their own, most of their movement is due to the turning of the Sun, *solar rotation.* You're repeating Galileo's discovery and doing it safely.

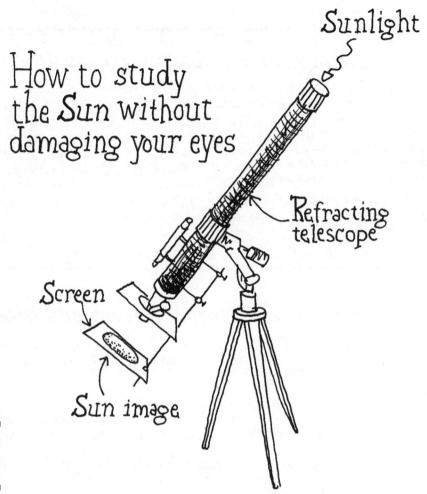

How to study
the Sun without
damaging your eyes

Sunlight

Refracting
telescope

Screen

Sun image

Figure 10-4:
Projecting
the Sun.

You must take great care in using the telescope as a solar image projector, and you must *never* allow an unsupervised child or any person not trained in this method to operate the telescope themselves. Don't look at the Sun through your telescope, and don't look at the Sun through the small finder telescope or viewfinder that your telescope may be equipped with. Make sure that no part of your body, anyone else's body, or any property gets in the projected beam of sunlight, only your cardboard screen.

If you don't want to use the projection technique or you have only more advanced telescopes (that use both lenses and mirrors) that should not be used with this technique, you can still view the Sun safely but you need a special solar filter. That process requires a significant investment, but it's well worth it. Read on, Galileo!

Seeing more when price is no object

Special solar filters called *H-alpha filters* enable you to see many more of the Sun's features than you can otherwise see in white light. In particular, these filters are wonderful for viewing *solar prominences,* which look like fiery arches on the edge or limb of the solar disk But these filters are very expensive (typically over $1,000).

If the price doesn't deter you, first get some experience in white-light solar observing, and then investigate H-alpha filters. Two manufacturers that sell them are Thousand Oaks Optical

(www.thousandoaksoptical.com) and the Coronado Instrument Group in Pearce, Arizona, whose Web site is at www.coronadofilters.com.

You may need a telescope adapter plate to hold one of these H-alpha filters, which are not necessarily made to mate easily with specific telescope models.

Seeing the Sun through front-end filters

The only solar filters that I recommend are those that go *at the front end of your telescope,* so that no light can enter the telescope without passing through the filter.

Filters that are placed at, near, or in place of the eyepiece can in many cases break as a result of concentrated solar heat, with possible extreme danger to your vision. Use only filters that go at the front end of your telescope.

The following are telescope filters for the front end of your telescope that are recommended for use when you are viewing the Sun:

- ✔ **Full-aperture filters:** Appropriate for telescopes of 4-inch aperture or less (the *aperture* is the diameter of the light-collecting mirror or lens in your telescope), such as the Meade ETX-90/EC. The filter extends across the full diameter of the telescope, so that the entire light-collecting mirror or lens receives the filtered light from the Sun.

- ✔ **Off-axis filters:** Best for telescopes of 4-inch aperture or more that are not refractors. An off-axis filter is smaller than the aperture of the telescope, but it's mounted in a plate that covers the entire aperture. The Sun is so bright that you don't need the whole aperture of the telescope to collect enough light for good solar viewing. It's true that a larger aperture can also give a sharper view, but in most viewing locations, blurring by Earth's atmosphere makes it impossible to obtain a solar image as sharp as the whole aperture of a 4-inch or greater telescope would provide. The less unneeded sunlight that gets into your telescope, the safer you and the telescope will be.

Stopping down

When you block some or most of the light path of a telescope (for example, by using a filter that admits light only across part of the aperture), that's called *stopping down the telescope.* Tell somebody at the astronomy club that you've been watching the Sun "with my telescope stopped down" and they'll think that you're a pro!

Guess who invented the stopping down of telescopes? Galileo! What a guy! You can repeat his work by watching sunspots with a stopped-down telescope. But he also did physics experiments such as dropping weights from the Leaning Tower of Pisa. Don't even think about repeating that one. Stop before you drop.

You want an off-axis solar filter with most telescopes other than refractors because nonrefractors usually have little mirrors or mechanical devices that are on-center inside the telescope tube, blocking the part of the light that comes down the center of the tube.

In the special case of a refractor of 4-inch aperture or more, meaning a pretty expensive telescope, the filter used should go over the top end of the telescope, be smaller than the telescope aperture, but be mounted centrally in the plate that covers the telescope. The filter should be mounted on-center because, generally speaking, the central part of the primary or objective lens of the telescope (the big lens) may have better optical quality than the periphery of the lens.

You can obtain solar filters from a variety of sources. Here are two suppliers with a reputation for good quality:

✔ **Roger W. Tuthill, Inc.,** in Mountainside, New Jersey, sells trademarked Solar Skreen Sun Filters for telescopes, binoculars, cameras, and video camcorders of various kinds, including filters made specifically for many popular telescope models in the Celestron and Meade product lines. These filters are made of two sheets of special Mylar that are coated with aluminum. When you get the filter, it may look like it hasn't been stretched tight, but that's how this material behaves. And it works fine.

The filters for specific telescopes and other equipment come mounted in cells or holders that fit over your telescope or lens. But Tuthill also sells unmounted squares of Solar Skreen. On a recent eclipse cruise, I wrapped one of these squares around each of the large lenses of my binoculars, using rubber bands to hold them tight. I had great views of the eclipse from my deck chair while sipping piña coladas.

Only use Solar Skreen or any other solar filter in accord with the manufacturer's directions. See Tuthill's Web site at www.tuthillscopes.com.

✔ **Thousand Oaks Optical,** in Thousand Oaks, California, manufactures full-aperture and off-axis glass solar filters under the designation Type 2 Plus. These filters are good for viewing through your telescope.

The Thousand Oaks Type 3 Plus filters are used for photographing the Sun through telescopes, but are not dark enough for use in telescopic viewing of the Sun.

Thousand Oaks also sells Polymer Plus filters based on a plastic polymer film. It's their answer to Tuthill's Solar Skreen. Naturally, each manufacturer thinks that its product is best. See the Thousand Oaks Web site at www.thousandoaksoptical.com.

Having Fun with the Sun: Solar Observation

The Sun is a fascinating, constantly changing ball of hot gases that offers a lot to see for the prudent observer. With the proper precautions (see the preceding section), you can get set up to see for yourself. In addition to viewing the Sun by the projection method or by using telescopes equipped with solar filters, you can also visit Web sites that offer awe-inspiring, professionally produced pictures. This section suggests some ways that you can personally enjoy old Sol.

Tracking sunspots

After you become confident in your ability to observe the Sun safely by using the projection method, or equipping your telescope with a safe solar filter, you can begin your observation by tracking sunspots, using the following plan:

- Observe the Sun as often as possible (tell the boss you didn't oversleep; you were counting sunspots, not counting sheep).

- Note the sizes and positions of sunspots and groups of sunspots on the solar disk (the visible surface of the Sun that is facing toward Earth).

  Some sunspots look just like tiny dark spots. If the sunspot is truly a tiny dark spot, as seen even with a powerful observatory telescope, it's called a pore. But if a sunspot is big enough, you'll be able to distinguish its different regions. The dark central portion is called the umbra, and the surrounding area that's darker than the solar disk but lighter than the umbra is the penumbra.

- Chart the motion of the sunspots as the Sun makes one complete turn — which is every 25 days (at the equator) to about 35 days near the poles (yes, the Sun turns at different rates at different latitudes, one of its many mysterious and unexpected properties).

Getting your personal sunspot number

Compute your own *sunspot number* for each day of observation, using the formula:

$$R = 10g + s$$

R is your personal sunspot number, *g* is the number of groups of sunspots that you saw on the Sun, and *s* is the total number of sunspots that you counted, including those in groups. Sunspots usually appear isolated from each other on different parts of the solar disk. Those that are close together on one part of the disk are a group. And a spot that is all by itself counts as its own group (the reasoning behind this designation may be pretty spotty, but that's the way it's been done for many years).

Suppose that you discern five sunspots; three are close together in one place on the Sun, and the other two are at two widely separated locations. You have three groups (the group of three and the two groups consisting each of one spot), so *g* is 3. And the number of individual spots is 5. Then

$$R = 10 \times 3 + 5$$
$$R = 30 + 5$$
$$R = 35$$

Finding official sunspot numbers

On the same day, different observers come up with different personal sunspot numbers. If you have better viewing conditions and a better telescope, or just a better imagination, your sunspot number will be higher than the Jones's sunspot number. You just got R = 35, and that bum Jones could only account for R = 22. On sunspot numbers, you're not only keeping up with Jones, you're way ahead! "R" you clear on this?

Central authorities who tabulate and average together the reports from many different observatories find by experience that some observers keep up with the Joneses, some can't see as many, and some are far ahead. From this experience, they calibrate each observatory or observer and make allowances in future counts so they can average the reports and get the best estimate of the sunspot number for each day.

You can check out the sunspot number, as determined professionally, at `www.sunspot.noao.edu/IMAGES/sunspot_numbers.html`, a National Solar Observatory Web site.

The current sunspot cycle is expected to peak sometime around the year 2000. If you start watching spots before then, you may be able to judge the maximum for yourself, although the official decision is based on a complicated averaging scheme. And you can watch the number of spots decline over subsequent years until we approach the minimum of the sunspot cycle, when you might look in vain for a decent sunspot for months at a time.

Looking at solar pictures on the World Wide Web

You can see a current or recent professional photographs of the solar disk and sunspots (what solar astronomers call a white-light photograph — white light being all the visible light of the Sun). A good place to look is at the site of Italy's Catania Astrophysical Observatory (www.ct.astro.it/ sunoacf.html). The Catania observers call the sunspot number the *Wolf number,* after a famous solar astronomer, and they tabulate their numerical counts of groups and spots alongside their photographs. You can get experience in identifying a group and counting sunspots by counting the spots on their photos.

Sometimes, the weather is cloudy in Italy, so you need to look elsewhere for a professional white-light photo of the whole solar disk. A good place is the Web site of the Learmonth Solar Observatory in Western Australia. They enjoy lots of clear days, and they display a variety of solar images (www.ips.oz.au/ learmonth/solar/index.html). If the term "white light" is not clearly stated, look for an image that has *GONGWL* in the name. That's a white-light Sun photo taken as part of solar astronomers' worldwide GONG project, which is ringing in with new findings on the Sun at regular intervals.

When you become an advanced astronomer and are ready to photograph celestial scenes through your telescope, you may want to try solar photography, too. Examples to inspire you were taken at the Mount Wilson Observatory, which has been photographing the Sun since 1905. Check out the fantastic picture of an airplane silhouetted against a spotty Sun, and also their picture of the largest sunspot group ever photographed, from April 7, 1947. Should you be lucky enough to see a sunspot group even half as large as that one, it will probably be visible through a solar filter, without the need for a telescope. The Mount Wilson site for white-light solar photographs is at http://physics.usc.edu/solar/direct.html.

In fact, astronomers are studying the Sun in all kinds of light, not just white light. This research includes pictures taken in ultraviolet and extreme ultraviolet radiation and X-rays, which are all forms of light that are invisible to the eye and are, in fact, blocked by Earth's atmosphere. Those pictures are made with telescopes mounted on satellites orbiting Earth at high altitude, or spacecraft located farther away and orbiting the Sun just like Earth does. Sun images from satellites and from many kinds of telescopes on the ground are available on NASA's "Current solar images" Web site at (http://umbra.nascom.nasa.gov/ images/latest.html).

If your computer is equipped to view movies over the World Wide Web, you can see selected videos of the changing face of the Sun from the SOHO satellite (as long as it remains in operation) at NASA's SOHO Movie Theater site (http://sohowww.nascom.nasa.gov/synoptic/soho_movie.html).

SOHO, the Solar and Heliospheric Observatory, is a spacecraft built by the European Space Agency and loaded with scientific instruments, half of them supplied by NASA. So if you are from the United States or from western Europe, you've probably paid taxes to support this project. And even if you're from some other nation or you don't pay taxes, you can see the pictures, too.

Experiencing a total eclipse of the Sun

On a daily basis, the best way to see the Sun's outermost, most changeable, and most beautiful region, the corona, is in the satellite images posted on the Web sites in the preceding section

But seeing the corona "live and in person" is a spectacle that you should not deny yourself. It's one of nature's most beautiful sights. That's why many amateur astronomers save their earnings for years in order to splurge on a great eclipse trip. And professional astronomers find ways to get to the eclipse, too, even though they have their satellites and space telescopes.

There are *partial, annual*, and *total eclipses* of the Sun (see Figure 10-5). The great spectacle is the total eclipse; some annular eclipses are well worth the trip, too. During an annular eclipse, a thin bright ring of the photosphere is visible around the edge of the Moon. A partial eclipse is not something I'd go hundreds of miles out of my way to see, because you don't see the chromosphere or corona, but it's well worth practicing on if one comes your way. After all, the first and last stages of a total eclipse or an annular eclipse are partial eclipses! So you want to know how to observe those stages, too.

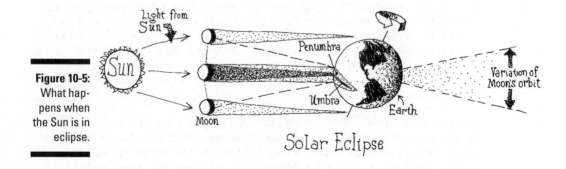

Figure 10-5: What happens when the Sun is in eclipse.

Observing a total eclipse

To observe a partial eclipse, or the partial eclipse phases of a total eclipse of the Sun, use Solar Skreen or the other solar filters I describe in the preceding section You can watch through binoculars or telescopes equipped with such a filter, or through a filter held in front of your eyes.

A total eclipse normally starts with a partial phase, beginning with *first contact,* when the edge of the Moon first comes across the edge of the Sun. Now the viewer sees a *partial eclipse* of the Sun, signifying that he or she is located in the *penumbra* or light outer shadow of the Moon. At *second contact,* the Moon's leading edge has reached the far edge of the Sun, so the Sun is totally blocked. Now you can witness *total eclipse;* you are in the dark *umbra* or central shadow of the Moon. This period is when you can put down your viewing filter or your filtered binoculars and just stare safely at the fantastic sight of the totally eclipsed Sun. But you must not stare at the Sun when totality is over.

The corona forms a bright white halo around the Moon, perhaps with long streamers extending east and west. Thin bright polar rays may be seen off the north and south limbs of the Moon and all around the lunar limb. Watch for small bright red points, which are solar prominences visible to the naked eye during brief moments of the eclipse. Near the peak of the 11-year sunspot cycle, the corona is often round, but near sunspot minimum, it's elongated east-west. The corona's shape is different at every eclipse.

Some people take the solar filters off their binoculars and telescopes and look at the totally eclipsed Sun through these instruments without benefit of filters. This approach is very dangerous if

- ✔ You look too soon, before the Sun is really totally eclipsed.

- ✔ You look too long (a very easy way to have an accident), and continue to watch through an optical instrument after the Sun begins to emerge from behind the Moon.

Be warned! I strongly advise against telescopic and binocular viewing of the Sun without filters even during total eclipse, unless you're under the direct control of an expert. Sometimes, for example, the experienced leader of an eclipse trip or cruise group uses a public address system, computer calculations, and personal observing know-how to announce when you can look at the eclipsed Sun this way and when you must stop, giving the audience plenty of warning.

In my experience (which is painful), the easiest way to hurt yourself is to linger looking through your binoculars or telescope "just another second or two" when a tiny part of the bright visible surface of the Sun has started to emerge from behind the Moon. That tiny bright part may not make you immediately avert your vision, because it doesn't seem brilliant enough. But what you don't realize is that the infrared rays from that little exposed part of the solar surface are damaging your eye without dazzling it or causing immediate pain. In a few minutes or less, you will begin to feel the pain. But, then, the damage is done.

So bright, you have to wear shades

A company named Rainbow Symphony, Inc., of Reseda, California, is a prominent supplier of solar filters mounted in inexpensive eyeglass frames, like the 3-D viewers used for certain movies. (They sell 3-D viewers, too, but they won't do you any good at an eclipse.) Their product goes under the name Eclipse Shades. These items are sufficiently inexpensive that I suggest bringing one pair of Shades for each person in your party, even though you may purchase more expensive solar filters for your optical instruments. Usually, the organizers of tours and eclipse cruises distribute viewers, but sometimes they cut corners and hand out strips of aluminized Mylar, which work, but are less convenient than the mounted Shades. See the Rainbow Symphony Web site at www.rainbowsymphony.com/soleclipse.html.

Be safe, follow all directions, never take chances looking at the Sun, and you can look forward to many happy returns of total eclipses of the Sun!

Seeking shadow bands and Baily's Beads

Another good reason to avoid looking at the Sun with optical instruments during the total phase of an eclipse is that there's so much to look for all around the sky with your naked eye.

- ✔ Just before totality, so called *shadow bands*, shimmering low-contrast patterns of dark and light stripes, may race across Earth or across the deck of your ship. They are an optical effect produced in Earth's atmosphere when the bright disk of the Sun is reduced to the last little sliver by the eclipsing Moon, but is not yet completely eclipsed.

- ✔ *Baily's Beads* are another wonderful and fleeting sight during a total eclipse. They occur just instants before and after totality, when little regions of the bright solar surface shine through between the mountains or crater rims on the edge of the Moon.

- ✔ And observe the wildlife (and domesticated animals if you are in such an area). Birds come down to roost, cows may head back for the barn, and so on. At one 19th-century eclipse, some top scientists set up their instruments in a barn, pointing the telescopes out through the door. Boy were they surprised when totality began and the livestock ran in!

When the Sun's totally eclipsed, look at the dark sky all around the Sun. This is a rare chance to see stars in the daytime. Special articles published in astronomy magazines, or posted on their Web sites, will tell you which stars and planets to look for. Or you can figure it out yourself by simulating the date and time of the eclipse on your desktop planetarium, setting the program to display the sky as it will be from the place where you expect to be observing.

Following the path of totality

Totality ends at *third contact,* when the Moon's trailing edge moves out across the solar disk. Now you are back in the penumbra, and you can see a partial eclipse. At *fourth* or *last contact,* the Moon's trailing edge moves off the forward limb of the Sun. The eclipse is over.

The whole eclipse, from first contact to last contact, may take a few hours, but the good part, totality, lasts from less than a minute to seven minutes, or slightly more.

And there's only one place on the *path of totality* — the track of the center of the Moon's shadow across the surface of Earth — where the duration of totality is at its greatest. Everywhere else on the path, totality is briefer. Of course, the place where the eclipse has maximum duration may not be the place where the weather prospects are best, or it may not be a place that's easily or safely reached. So advance planning of your eclipse trip is vital. At any good location, all the accommodations, rental vehicles, and so on, will be booked up at least a year or two in advance of the eclipse.

To plan your eclipse trip, pick a likely eclipse from Table 10-1, and start investigating the best way to view it.

Table 10-1	Coming Total Eclipses of the Sun	
Date of Total Eclipse	**Maximum Duration (Minutes and Seconds)**	**Path of Totality**
June 21, 2001	4:57	Across southern Atlantic and Africa from Angola to Mozambique and then over the Indian Ocean and Madagascar
Dec. 4, 2002	2:04	Eastern Atlantic, across Africa from Angola to Mozambique, across the Indian Ocean to South Australia
Nov. 23, 2003	1:57	Antarctica and the southern Indian Ocean
Apr. 8, 2005	0:42	South Pacific to Central America and on to Venezuela
Mar. 29, 2006	4:07	Eastern Brazil across the Atlantic to Ghana, across Africa to Libya, across the Mediterranean Sea to Turkey, across the Black Sea to Georgia and Kazakhstan

(continued)

Table 10-1 *(continued)*

Date of Total Eclipse	Maximum Duration (Minutes and Seconds)	Path of Totality
Aug. 1, 2008	2:27	Northern Canada, Greenland, Arctic Ocean, Russia, Mongolia, to China
July 22, 2009	6:39	India, Nepal, Bhutan, China, across the China Sea and the central Pacific
July 11, 2010	5:20	Across the south Pacific and Easter Island to southern Chile and Argentina
Nov. 13, 2012	4:02	Australia, across the south Pacific toward (but not reaching) Chile
Nov. 3, 2013	1:40	Across the Atlantic to Africa from Gabon to Uganda, Kenya, and Ethiopia
Mar. 20, 2015	2:47	Across the north Atlantic south of Greenland to the Faeroe Islands, Norwegian Sea, Spitsbergen Island, and on almost to the North Pole

A few years beforehand, articles with information on the weather prospects and logistics for viewing from various locations will begin to appear in the astronomy magazines. Check the *Sky & Telescope* and *Astronomy* magazine Web sites. Look for eclipse tour advertisements in the magazines and on the Web. Check out the most reliable eclipse predictions on the NASA eclipse site at sunearth.gsfc.nasa.gov/eclipse.

Have a great time!

Chapter 11

The Stars: Nuclear Reactors

In This Chapter

▶ Following the life cycles of the stars

▶ Understanding star types

▶ Finding binary and variable stars

▶ Stargazing

▶ Meeting stellar personalities

Hundreds of billions of stars, like the Sun, make up the Milky Way galaxy, where Earth resides. Likewise, the billions of other galaxies that can be found in the universe all contain huge numbers of stars. And just like people, stars fit into dozens of classifications, but the overwhelming majority can be reduced to several, simple types, These types correspond to stages in the life cycles of stars, just as people can be classified by their ages.

Once you understand what a star is and how it runs through its life cycle, you'll have an overall feel for these shining beacons of the night sky, and those that aren't so bright, too.

In this chapter, I emphasize the initial mass (or size) of a star — what it's born with — as the main determinant of what it will become. Then I continue with the key properties of stars, and the features of binary and variable stars that make them so interesting for you to observe.

And no discussion of stars would be complete without some "gossip" about the celebrities. So I introduce you to some luminaries of the night sky that you'll want know — the leading "personalities" of the solar neighborhood.

Life Cycles of the Hot and Massive

The most important star categories correspond to successive stages in their life cycles: babies, adults, seniors, and the dying. (What! No teenagers? The universe gave up on youth classifications after the terrible twos!) Of course, no astrophysicists worth their Ph.D.s would use such simple terms, so astronomers refer to these types of stars as young stellar objects (YSOs), main sequence stars, red giants, and those in the end states of stellar evolution, respectively. (You'll be glad to know that no star ever dies completely; at most, it "evolves" into a new and final state such as a white dwarf or a black hole.)

Here's the life cycle of a normal star with about the same mass as the Sun:

1. The star is created when gas and dust in a cool nebula condenses, forming a young stellar object (YSO).

2. Shrinking, the star dispels its remaining birth cloud, and its hydrogen fire ignites. In other words, nuclear fusion is underway, as I explain in Chapter 10.

3. As the hydrogen burns steadily, the star joins the main sequence (a stage in stellar life that I describe later in this section).

4. When the star uses up all the hydrogen in its core, the hydrogen in the shell (a larger region surrounding the core) ignites.

5. The energy released by the burning of the hydrogen shell makes the star brighter and expands it, making its surface larger, cooler, and redder, so it becomes a so-called red giant star.

6. Stellar winds blowing off the star gradually expel its outer layers, which form a planetary nebula around the remaining hot stellar core.

7. The nebula expands and dissipates into space, leaving just the hot little core.

8. The core, now a white dwarf star, cools and fades forever.

Stars with much higher masses than the Sun have different life cycles; instead of producing planetary nebulae and dying as white dwarfs, they explode as supernovas and leave behind neutron stars or black holes. And this cycle happens rapidly; the Sun may last 10 billion years, but a star that begins with 20 or 30 times the Sun's mass will explode just a few million years after it is born.

Stars with much lower masses than the Sun hardly have a life cycle at all. They begin as YSOs, and then join the main sequence and remain there as red dwarfs forever. The explanation for this is a fundamental principle of stellar astrophysics: the bigger the mass, the fiercer and faster the nuclear fires burn, so the smaller the mass, the less fiercely it burns and the longer it lasts.

By the time the Sun uses up its core hydrogen, it will be at least nine billion years old. But a red dwarf star burns hydrogen so slowly that it goes on that way forever (for all practical purposes).

The following sections describe the stages in stellar lives in more detail.

YSOs: Taking tiny star-steps

Young stellar objects (YSOs) are newborn stars that are still surrounded by, or are trailing, wisps of their birth clouds. They include *T Tauri stars,* named for the first one of their type: the star "T" in constellation Taurus; and *Herbig-Haro objects,* named for the two astronomers who classified them. (Actually, H-H objects are glowing blobs of gas expelled in opposite directions from the young star itself, which is usually hidden from view by dust from its birth cloud.) YSO's can be found in stellar nurseries, called HII regions by astronomers, such as the Orion Nebula (see Figure 11-1), where hundreds of stars have been born in the past one or two million years.

Figure 11-1: The Orion Nebula is among those nebulae where many stars are born, at first shrouded behind clumps of interstellar dust.

Jerry Lodriguss

Many of those Hubble Space Telescope images of spectacular jet-like nebulae are pictures of YSOs. The jets and other nebular surroundings are prominent, but the stars themselves are sometimes barely visible (if they can be seen at all), hidden by the gas and dust around them.

Main sequence stars: A long adulthood

Main sequence stars, which include the Sun, have shed their birth clouds and are shining, thanks to the nuclear fusion of hydrogen into helium that goes on in their cores (see Chapter 10 for more about nuclear fusion in the sun). For historical reasons, going back to when astronomers classified stars before they understood their differences, main sequence stars are also called dwarfs (never "dwarves"). A main sequence star is a dwarf even if it's ten times more massive than the Sun or even much larger.

When astronomers and newspaper science writers refer to "normal stars," they often mean main sequence stars. When they write about "sunlike stars," they mean main sequence stars with roughly the same mass as the Sun, give or take a factor of no more than two.

The smallest main sequence stars are *red dwarfs*, which shine with a dull red glow.

Red dwarfs have little mass, but there's a mess of them. The vast majority of main sequence stars are red dwarfs. They are like those tiny gnats at the seashore that folks call "no-see-ums"; they are all around us, but you can hardly spot them. Red dwarfs are so dim that even the nearest one, Proxima Centauri — which is, in fact, the nearest-known star beyond the Sun, can't be seen without telescopic aid.

Red giants

Red giant stars are another kettle of starfish entirely. Red giants are much larger than the Sun. Often, they're as big around at the equator as the orbit of Venus or even the orbit of Earth. They represent a later stage in the life of an *intermediate-mass* star — one with several times more to somewhat less than the mass of the Sun — after it has graduated from the main sequence category.

A red giant is not burning hydrogen in its core. In fact, it's burning hydrogen in a spherical region just outside the core, called a *hydrogen-burning shell.* It can't burn hydrogen in its core because it has already burned up all its core hydrogen, turning it to helium by nuclear fusion. Stars that are much more massive than the Sun don't become red giants; they swell up so much we call them *red supergiants.* A typical red supergiant can be one or two thousand times larger than the Sun, and big enough to extend past the orbit of Jupiter, or even Saturn, if put in the Sun's place.

Stars in the end states of stellar evolution

End states of stellar evolution is a polite, catchall term for stars whose best years are far behind them. The term encompasses

- White dwarfs
- Central stars of planetary nebulae
- Neutron stars
- Supernovas
- Black holes

They all are stars on their final glide paths, dying or doomed to oblivion.

The bigger, the rarer

SETI observers (see Chapter 14) don't point their radio telescopes at massive stars to search for radio signals from advanced civilizations because massive stars explode and die after lifetimes so short that it's difficult to imagine life evolving on any surrounding planets before the end comes, let alone having enough time to float a bond issue for planetary-sized transmitters.

Massive stars are much rarer than low-mass stars. The more massive the stars, the fewer they are. So eventually, as existing stars age and the birth clouds for new stars are used up, the Milky Way will consist overwhelmingly of just two types of stars. They will be the red dwarfs that go on more or less forever, and the white dwarfs that do about the same, but fade as they go. Yes, there will be lots of neutron stars and stellar mass black holes, but because they are the remains of more massive stars, they will be numerically insignificant compared to the red dwarfs and white dwarfs, which come from the most abundant types of main sequence stars.

Stars are like people in that the biggest ones are rare, just as 7'4" people like center Rik Smits of the Indiana Pacers basketball team are few and far between.

White dwarfs

White dwarfs can actually be blue, white, yellow, or even red, depending on how hot they are. They are the remains of sunlike stars and take after the old generals who, according to Douglas MacArthur, never die — they just fade away.

A white dwarf is like a glowing coal from a fire that you have just extinguished. It's not burning any more, but it's still hot. It will fade away over all eternity as it cools down. White dwarfs are compact stars — small and very dense. A typical white dwarf may have as much mass as the Sun, yet be little bigger (if at all) than Earth. White dwarfs are no-see-ums, too; they are the most common stars after red dwarfs, but even the closest white dwarf to Earth is too dim to be seen without a telescope.

So much matter is packed into such a small space in a white dwarf star that a teaspoon of it would weigh about a ton on Earth. Don't try measuring it with your good silver; the spoon will be all bent out of shape.

According to a leading college textbook, *The Cosmic Perspective* by Jeffrey Bennett, Megan Donahue, Nicholas Schneider, and Mark Voit (Addison-Wesley Publishing Company, 1999), "Two dice made from white dwarf material would weigh five tons — about as much as three cars." Try rolling those dice at Las Vegas!

Central stars of planetary nebulae

Central stars of planetary nebulae are little stars at (duh) the centers of small, beautiful nebulae, such as the famous Ring Nebula in Lyra, pictured in the color section of this book.

Central stars of planetary nebulae are a lot like white dwarfs and in fact turn into them, if they are not white dwarfs already. So they, too, are the remains of sunlike stars. The nebulae, each composed of gas that was expelled from a star over tens of thousands of years, expand and fade and blow away and eventually they leave behind stars that are no longer the centers of anything — they are just white dwarfs.

Neutron stars

Neutron stars are so small that they look up to white dwarfs, but they also outweigh them. (More accurately, they "outmass" them. Weight is just the force a planet or other body exerts on an object of a given mass. You would weigh different amounts on the Moon, Mars, or Jupiter than you do on Earth, even though your mass would stay the same.)

Neutron stars are like Napoleon, small in stature but not to be underestimated. A typical neutron star is only one or two dozen miles across, but has half again or even twice the mass of the Sun. A teaspoon of neutron star would weigh about a billion tons on Earth. *The Cosmic Perspective* authors calculate that "A paper clip made from neutron star material would outweigh Mount Everest."

Some neutron stars are better known as *pulsars*. Figure 11-2 shows the Crab Nebula, which has a pulsar at its center.

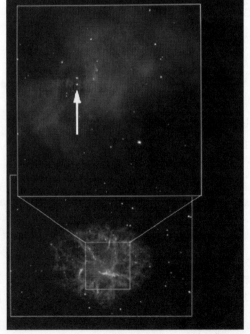

Figure 11-2:
A Hubble close-up at top, shows the pulsar (arrow) in the Crab Nebula.

Courtesy of NASA

A pulsar is a neutron star that is highly magnetized, spinning rapidly, and producing one or more beams of radiation (which may consist of radio waves, X-rays, gamma rays, and/or visible light). As a beam sweeps past Earth like a searchlight from a galactic supermarket opening, our telescopes receive brief spurts of this radiation, which we call pulses. You can guess from that how pulsars got their name. Your pulse rate tells how rapidly your heart is beating. A pulsar's rate tells how fast it is spinning. The rate can be a few hundred times per second, or just once every few seconds.

Supernovas

Supernovas (which professional experts call *supernovae,* as though they all know Latin like old-time scientists) are enormous explosions that destroy entire stars (see Figure 11-3).

Figure 11-3:
A supernova
in the spiral
galaxy M51.

Supernova Near Nucleus of Galaxy M51

Hubble Space Telescope · Wide Field Planetary Camera 2

Courtesy of NASA

The first type you need to know about is Type II (hey, I didn't invent the numbering system). A *Type II supernova* is the brilliant, catastrophic explosion of a star much more massive, larger, and brighter than the Sun. Before exploding, it was a red supergiant star, and maybe even hot enough to be called a blue supergiant. Regardless of color, when a supergiant explodes, it may leave behind a little souvenir, which is a neutron star. Or much of the star may implode (fall in its own center) so effectively that it leaves behind an even weirder object, a black hole.

The second type of supernova that is especially important is called a Type Ia. *Type Ia supernovas* are even brighter than Type II, and they explode in a reliable manner. When astronomers observe a Type Ia supernova, we can figure out how far away it is by how bright it looks. The farther away it is, the dimmer the supernova looks. Type Ia supernovas are used by astronomers to measure the universe and how it is expanding. In 1998, two groups of astronomers studying Type Ia supernovas discovered that the expansion of the universe is not slowing down — it is getting faster. This discovery led experts to revise their theories of cosmology and the Big Bang (see Chapter 16).

Type Ia supernovas all produce similar explosions because they are eruptions in binary systems in which gas from one star flows down onto the other (a white dwarf), building up an outer hot layer that reaches a kind of critical mass and then explodes, shattering the star. With less than critical mass, no explosion occurs; with critical mass, a standard explosion results, with more than the critical mass . . .wait — you can't have more than critical mass because the star will have exploded! Astrophysics isn't so hard.

Black holes

Black holes are objects so dense and compact that they make neutron stars and white dwarfs feel like cotton candy. So much matter is packed into such a small space in a black hole that its gravity is strong enough to prevent anything, even a ray of light, from getting out. Physicists consider that the contents of a black hole have left our universe. If you fall into a black hole, you can kiss your universe good-bye.

You can't see the light from a black hole because the light can't get out, but scientists can detect black holes by their effects on their surroundings. Matter in the vicinity of a black hole gets hot and rushes madly around, but never gets organized. Instead, it falls into the black hole and "that's all folks." This is all due to the powerful gravity of the black hole.

Actually, I over simplified; a bit of the matter swirling around the black hole does escape, just in time, sometimes. It gets shot out in powerful jets at a significant fraction of the speed of light (which is 186,000 miles per second, in a vacuum such as outer space).

That's how scientists detect black holes: We see gas swirling around them that's just too hot for normal conditions, we detect jets of high energy particles making their escape and avoiding falling into the black hole, and we even detect stars racing around orbits at fantastic speeds, as though they are driven by the gravitational pull of an enormous unseen mass (which they are).

Up until April 1999, when astronomers announced that they had discovered a third class of black holes — intermediate black holes — two types of black holes were recognized:

- Stellar mass black holes
- Supermassive black holes

A *stellar mass black hole* has — you guessed it again — the mass of a star. More accurately, these black holes range from about three times the mass of the Sun up to perhaps 100 times the solar mass, although none as heavy as that has been found. These black holes are about the size of neutron stars. A black hole with a solar mass of 10 has a diameter of about 37 miles (60 kilometers). If you could squeeze the Sun down to a sufficiently compact size that it became a black hole (fortunately, this is probably impossible), its diameter would be 3.7 miles (6 kilometers). Stellar mass black holes form in supernova explosions and possibly by other means.

A *supermassive black hole* has a mass of hundreds of thousands to even a few billion times the mass of the Sun. Generally, they are found at the center of galaxies. Our own Milky Way has a central black hole, known as Sagittarius A* (nope, that's not an asterisk to refer you to a footnote: this name is really pronounced "Sagittarius A star"). It weighs in at about one million solar masses, and we in the solar system orbit around that black hole once every 226 million years. That's the latest value from the Very Long Baseline Array, a radio telescope with component antennas that stretch across U.S. territory from the Virgin Islands, through North America, and out to Hawaii. Some astronomers think that a supermassive black hole is at the center of every galaxy or at least at the center of every full-sized galaxy. We're not so sure about dwarf galaxies. I explain more about supermassive black holes in Chapter 13.

Intermediate mass black holes got that clever name from the experts who discovered them but weren't sure what they were. Some scientists think they are just the teenage stage of a future supermassive black hole, much lighter than they will be some day, but swallowing everything in sight and destined for a great mass. Others say that they may be something else entirely, but if so what? Inquiring minds want to know, but right now, more research is needed. They have masses around 500 to 1000 times the mass of the Sun.

To tell the truth, supermassive black holes are not stars. And, most likely, neither are intermediate mass black holes. But I've got to mention them some place. You can't call yourself an astronomer if you don't know about black holes. Once you're ready to pass yourself off as an astronomer, folks will ask you all kinds of questions about black holes. But how many questions do you think that you will get about main sequence stars and young stellar objects?

Diagramming the Stars: Temperature, Mass, and Hertzsprung-Russell

The significance of the different types of stars becomes more clear when basic observational data are plotted on an astrophysicist's graph. The data are the magnitudes (or brightness levels) of the stars, which are plotted on the vertical axis, and the colors (or temperatures), which are plotted on the horizontal axis. The graph is called a *color-magnitude diagram,* also the Hertzsprung-Russell or H-R diagram, after the two astronomers who first made such an illustration (see Figure 11-4).

As an Astronomy 101 teacher at UCLA and the University of Maryland, I could always tell who studied and who didn't. When I asked on the mid-term exam, "what are plotted on the H-R diagram," those students who answered "H and R," were revealed to be in sore need of remedial reading.

Spectral types: What color is my star?

Hertzprung and Russell didn't have good information on the colors or temperatures of stars, so they plotted the spectral type on the horizontal axis of their original diagram. The *spectral type* is a parameter assigned to a star on the basis of its spectrum. And the *spectrum* is the way that the light of a star appears when it is spread out by a prism or other optical device in an instrument called a spectrograph.

At first astronomers didn't have a clue about what special types meant, so they just grouped stars together (under the headings of Type A, Type B, and so on) based on similarities in their spectra. Later, astronomers realized that the spectral types reflected the temperatures and other physical conditions in the atmospheres of the stars, where their light emerges into space. Once scientists understood what the colors signified, they could organize the spectral types in temperature order and Hertzsprung and Russell could plot them on their diagram. Some superfluous types were dropped.

The main spectral types on the H-R diagram are O, B, A, F, G, K, M, going from the hottest stars to the coolest ones. College students memorize this sequence with the help of mnemonic devices such as "Oh, be a fine girl (guy), kiss me."

Table 11-1 describes the general properties of stars in each of the spectral classes.

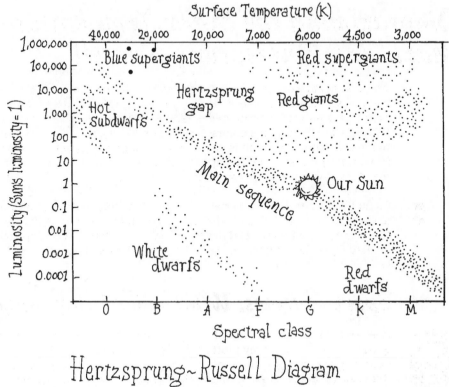

Figure 11-4:
Hertzsprung-
Russell
diagram.

Hertzsprung~Russell Diagram

Table 11-1		Spectral Classes of Stars	
Class	*Color*	*Surface Temperature*	*Example*
O	Violet-white	30,000° K and more	Lambda Orionis
B	Blue-white	12,000° K to 30,000° K	Rigel
A	White	8,000° K to 12,000° K	Sirius
F	Yellow-white	6,000° K to 8,000° K	Procyon
G	Yellower white	5,000° K to 6,000° K	Sun
K	Orange	3,000° K to 5,000° K	Arcturus
M	Red	<3,000° K	Antares

Star light, star bright: Classifying luminosity

Each spectral class has subdivisions. For example, the Sun has a G2V spectrum, meaning that it is a G-type star that is slightly cooler than a G0 or G1 star and slightly hotter than a G3 star, but much cooler than a K star, and that it is a main sequence dwarf, which is indicated by the "V." The "V" is called the Sun's luminosity class. Each star has a luminosity class, represented by a Roman numeral.

Supergiants are luminosity classes I and II, giants are class III, and subgiants (a minor stage between main sequence and red giants) are luminosity class IV. All red dwarfs are luminosity class V, and white dwarfs are class D.

Today, you can find H-R diagrams that look quite different from each other in format, but that all present the same data: the relative properties of the stars as indicated by their temperatures and brightness.

Some H-R diagrams are calibrated, so that they show the true brightnesses or luminosities of the stars, not the apparent magnitudes or brightnesses as seen from Earth.

Mass determines class

A star with greater mass concentrates a fiercer nuclear fire in its core and produces more energy than a star of lower mass. So the more massive main sequence star is brighter and hotter than the less massive main sequence star. The more massive stars are bigger, too. With this information, you can follow the fundamental point of stellar astrophysics reflected in the H-R diagram: Mass determines class.

On the H-R diagram (such as Figure 11-4) magnitude is plotted with brighter magnitudes (which have smaller numerical values) higher on the graph, and spectral class is plotted with the hotter stars to the left and the cooler ones to the right. So temperature runs from right to left, and magnitude runs from top to bottom.

On any H-R diagram, the plot of real observational data, where each point represents a single star, reveals a great deal to the careful reader:

  ✔ Many or most stars are on a band that runs diagonally from upper left to lower right. The diagonal band is the main sequence, and all the stars on the band are normal stars like the Sun, burning hydrogen in their cores.

✔ Some stars are on a wider, sparser, and roughly vertical band that stretches up and a little to the right from the diagonal band. The roughly vertical band that stretches up to the right (that is toward brighter magnitudes and cooler temperatures) is the giant sequence. It consists of red giant stars.

✔ A few stars are located all across the top of the diagram, from left to right. The small number of stars that are located across the top of the H-R diagram are supergiants; blue supergiants are on the left side of the diagram, more or less, and red supergiants (which outnumber the blue ones) are on the right side.

✔ A few stars are located far below the diagonal band, down at the bottom left and bottom center of the diagram. These stars are white dwarfs.

A main sequence star is plotted on the H-R diagram according to its brightness and temperature, but its brightness and temperature depend on only one thing: its mass. The diagonal shape of the main sequence represents a trend from high mass to low mass stars. The stars on the upper left of the main sequence have higher masses than the Sun, and the stars on the far right have lower masses.

Astronomers usually don't plot young stellar objects on the same H-R diagram with all the other stars, but if they did, the YSOs would be on the right side of the diagram, above the main sequence but not nearly as high up as the supergiant stars. Neutron stars and black holes are too dim to show up on the H-R diagrams where normal stars are plotted.

Interpreting the H-R diagram

With just a little more explanation, you too can be a stellar astrophysicist and understand at one fell swoop why all these stars fall into different parts of the diagram. Researchers spent decades figuring all this out, but you get it on a plate in *Astronomy For Dummies*. To keep it simple, I discuss a calibrated H-R diagram, where all the stars are plotted according to their true brightnesses.

Consider this: Why is one star brighter or dimmer than another? Two simple factors determine the brightness of a star: temperature and surface area. The bigger the star, the more surface area it has, and every square inch (or square centimeter) of the surface produces light. The more square inches, the more light. But what about the amount of light produced from a given square inch? Hot things are brighter than cool things, so the hotter the star, the more light it generates per square inch of surface area.

Got all that? Here's how it all goes together:

- ✔ **White dwarfs** are near the bottom of the diagram because they are very small. With very few square inches of surface area (compared to normal stars like the Sun), they just don't shine very brightly. As they fade away like old generals, they move down (because they are getting dimmer) and farther to the right (because they are getting cooler) in the H-R diagram. You don't see many of them on the right side of the H-R diagram because the cool white dwarfs are so faint that they usually fall below the bottom of the diagram as reproduced in books, and astronomers don't see and measure many of these faint ones.

- ✔ **Supergiants** are near the top of the H-R diagram because they are very big. A red supergiant can be more than 1,000 times larger than the Sun, so that if put in the Sun's place it would extend beyond the orbit of Jupiter. With all that surface area, supergiants are naturally very bright.

  The fact that the supergiants are roughly at the same heights on the diagram from left to right indicates that the blue supergiants (those on the left side) are smaller than the red supergiants (the ones on the right). How do we know that? They are blue because they are hotter, and if they are hotter, they produce more light per square inch. Because their magnitudes are nevertheless roughly the same (all the supergiants are near the top of the diagram), the red ones must have larger surface areas in order to produce equal total light with less light coming from each unit area than the blue ones.

- ✔ **Main sequence** stars are on the diagonal band that runs from upper left to lower right because the main sequence consists of all stars that burn hydrogen in their cores, regardless of their size. But differences in the sizes of the main sequence stars affect only where they appear on the H-R diagram. The hotter main sequence stars, those on the left side of the diagram, are also bigger than the cool main sequence stars. So hot main sequence stars have two things going for them: They have larger surface areas, and they produce more light per square inch than the cool stars. The main sequence stars at the far right are very dim and cool. They are the red dwarfs.

The main sequence is naturally in the middle of the H-R diagram, because all other stars are brighter or cooler (are higher up or lower down on the diagram) than main sequence stars.

Born Together, Stay Together: Binary and Multiple Stars

About half of all stars come in pairs. These *binary stars* are *coeval,* a fancy term for "born together." Stars that are born together, united by their mutual gravity as they condense from their birth cloud, usually stay together. What gravity unites, few celestial forces can break apart. A grown star in a binary system has never had any other partner.

A *binary star* consists of two stars that are each orbiting their common center of mass. The center of mass of two stars that have exactly equal masses is exactly halfway between them. But if one star has twice the mass of the other star, the center of mass is closer to the heavier star. In fact, it's twice as far from the lighter star as from the heavier star. If one star has one-third the mass of its heavy companion, it is three times farther from the center of mass, and so on. The two stars are like kids on a seesaw. The heavier kid has to sit closer to the pivot, so the two of them will be in balance.

The two stars in a binary system have orbits that are the same size if they have equal mass, and unequal size if they have unequal mass. The general rule is, the big guys follow smaller orbits. You might think that binary systems are like our solar system where the closer a planet is to the Sun, the faster it goes and the less time it needs to make one complete orbit. That would be a smart idea, but it would be wrong.

In binary systems, the big star that follows the smaller orbit travels more slowly than the little star in the big orbit. In fact, their respective speeds depend on their respective masses. The star that has one-third the mass of its companion moves three times as fast. By measuring these orbital velocities, astronomers can determine the relative masses of the members of a binary star.

Where two or more gather

Double stars are two stars that look very close to each other as seen from Earth. Some of them are true binaries, orbiting their common centers of mass. But others are just *optical doubles,* two stars that happen to be in nearly the same direction from Earth, but at very different distances. They have nothing to do with each other; they haven't even been introduced.

Triple stars are three stars that look close to each other and, like the members of a double star, may or may not be all that close. But a *triple star system,* like a binary system, consists of three stars that are held together by their mutual gravitation and that all orbit a common center of gravity.

A comparison to wedded (or unwedded) bliss may be in order. "Three's a crowd" is a common expression of the instability in most romantic arrangements when a third person becomes involved. The same is true of triple star systems: They actually consist of a close pair or binary system and a third star in a much bigger orbit. If all three stars were on close orbits, they would interact gravitationally in chaotic ways, and, before you knew it, the group would break up as at least one star flew away, never to return. The triple system is effectively a "binary star" where one member is actually a very close star-pair.

Quadruple stars are often "double-doubles," consisting of two close binary star systems that each revolve around the common center of mass of the four stars.

Multiple stars is the collective term for star systems larger than binaries: triples, quadruples, and more. At some point, the distinction between a large multiple star system and a small star cluster is blurred. One is essentially the same as the other.

The Doppler Effect and the importance of being binary

The fact that the orbital speeds of the member stars of a binary system depend on their masses is what makes binary stars attract the high interest of astronomers. We have lots of theories about the masses of different kinds of stars, but few good ways to check them. There may be 40 ways to leave your lover but we have few ways to weigh the stars. Fortunately, instead of having to throw up their hands and give up, astronomers can learn about star masses by studying binary systems and making use of a basic property of the physics of observed sources of light.

If one star is three times more massive than the other, it moves around its orbit in a binary system at one-third the orbital speed of its companion star. So all we have to do to learn their relative masses (meaning how much heavier one is than the other) is to measure their velocities. It's rarely possible to actually track the stars as they move, because most binary stars are so far away that we can't see them moving around their orbits. But even at great distances, we can receive the light from a binary star and study its spectrum. That spectrum may be the combined light of both stars in the binary.

Here's all you need to know about the Doppler Effect, named for Christian Doppler, the 19th-century Austrian physicist.

The frequency or wavelength of sound or light as detected by an observer changes depending on the speed of the emitting source with respect to the observer. For sound, the emitting source might be a train whistle. For light, it might be a star. (Higher frequency sounds are said to have higher pitch; a soprano has a higher pitch than a tenor. Higher frequency light waves have shorter wavelengths and are bluer, while lower frequency light waves have longer wavelengths and are redder.)

According to the Doppler Effect:

✔ If the source is moving toward you the frequency gets higher, so

- The pitch of the train whistle seems to be higher
- The light of the star seems to be bluer

✔ If the source is moving away, the frequency gets lower, so

- The whistle has a lower pitch
- The star looks redder

TECHNICAL STUFF

Stellar spectroscopy in a nutshell

Stellar spectroscopy is the analysis of the lines in the spectra of stars. It's by far the astronomers' most important tool for learning the physical nature of stars. Spectroscopy reveals:

- Radial velocities (motions toward or away from Earth) of stars

- Relative masses, orbital periods, and orbit sizes of stars in binary systems

- Surface gravities of stars

- Magnetic fields and their strengths on stars

- The chemical composition of stars (what atoms are present and in what states they exist)

- Sunspot cycles of stars (well, starspot cycles)

All this information comes from measuring the positions, widths, and strengths (how dark or bright they are) of the little dark (or sometimes, bright) lines in the spectra of stars. Scientists analyze them with the help of the Doppler Effect to learn how fast stars are going, how big their orbits are, and what their relative masses are. And there are other effects, called the *Zeeman Effect* and the *Stark Effect*, which affect the appearance of the spectral lines. Applying this knowledge, we can learn the strength of the magnetic field on the star from the Zeeman Effect and the density and surface gravity in the star's atmosphere from the Stark Effect. The very presence of particular spectral lines, each of which comes from a specific kind of atom that is absorbing (dark lines) or emitting (bright lines)

light in the atmosphere of a star, tells us about some of the chemical elements that are present and at what temperatures.

The spectral lines even tell us what condition or *ionization state* the atoms are in. Stars are so hot that the atoms of iron, for example, may be stripped of one or more of their electrons. This makes them iron ions. Each type of iron ion, depending on how many electrons it has lost, produces spectral lines with different characteristic patterns and positions in the spectrum. By comparing the spectra of stars recorded with telescopes with the spectra of chemical elements and ions as measured in laboratory experiments or calculated with computers, astronomers can analyze a star without ever coming within light-years of it.

In cool stellar gases, much of the iron has lost only one electron per atom, so it produces the spectrum of singly ionized iron. But in the very hot parts of stars, such as the million-degree corona of the Sun, iron may have lost ten electrons; it's in a high-ionization state, produces the corresponding pattern of spectral lines, and that pattern is a clear indicator that there is a very high temperature region on the star.

Certain parts of the Sun's spectrum change along with the coming and going of disturbed regions on the Sun, which peak about every eleven years. Similar changes occur in the spectra of other Sunlike stars. So astronomers can even tell the length of the sunspot cycle of a distant star by using spectroscopy, although the star is too far away to ever catch a glimpse of its sunspots.

The train whistle is the official example of the Doppler Effect as it has been taught to generations of sometimes unwilling high school and college students. But who listens to train whistles anymore?

A more familiar analogy is the way you feel the waves on the ocean as you zip around in a motorboat. As you ride away from the beach toward the direction where the waves are coming from, you feel the boat rocked rapidly by the chopping of the waves. But when you head back toward the beach, the rocking is much gentler. In the first case, you were moving toward the waves, meeting them before they would have gotten to you if you stood (or floated) still. So the frequency at which they struck the boat was greater than if the boat had been at rest.

The spectrum of a star contains some dark lines, places (wavelengths or colors of light) where the star does not produce as much light as at adjacent wavelengths. These lines are caused by the absorption of light by various particle atoms in the atmosphere of the star. They form recognizable patterns. And when the star is moving back and forth, the Doppler Effect makes these patterns of lines move back and forth in the spectrum.

So by observing the spectra of binary stars and seeing how their spectral lines shift from red to blue to red again as the stars go around their orbits, astronomers can tell how fast they are going and what their relative masses are. And by seeing how long it takes for a spectral line to go as far to the red as it goes and then how long it takes to go as far to the blue it goes, to back to the red again, we can tell the duration or period of the binary star orbit.

If you know that the period of one complete orbit is 60 days, for example, and if you know how fast the star is going, then you can figure out the circumference of the orbit, and thus its radius. After all, if you drive nonstop from New York City to some town in upstate New York in three hours (good luck with the traffic!) at 60 miles per hour, you know that the distance traveled is three times 60, or 180 miles.

Variable Stars

Not every star is, as Shakespeare wrote, as "constant as the Northern Star." In fact, the North Star isn't constant either. It's a variable star, meaning one whose brightness changes from time to time. For many years, astronomers thought that they had the North Star's brightness changes down pat. It seemed to brighten a little and fade a little over and over, reproducibly. Then its expected changes, well, changed. This difference in the pattern could signify a physical change over time, and scientists are studying what it means.

Variable stars come in two basic types:

✔ *Intrinsic variable stars* are those whose brightness changes are produced by physical changes in the stars themselves. Intrinsic variable stars can be divided into three main categories:

- Pulsating stars

- Flare stars

- Exploding stars

✔ *Extrinsic variable stars* are those whose brightness changes because something outside the star alters its light, as visible from Earth. Two main types of extrinsic variable stars are:

- Eclipsing binaries

- Microlensing event stars

Pulsating stars: Everybody's favorites

Pulsating stars bulge in and out, getting bigger and smaller, hotter and cooler, brighter and dimmer. These stars are in a physical condition where they simply oscillate like throbbing hearts in the sky.

Cepheid variable stars

The most important pulsating stars, from a scientific standpoint, are the Cepheid variable stars, named for the first of their type to be studied, the star Delta in constellation Cepheus (Delta Cephei).

The American astronomer Henrietta Leavitt discovered that Cepheids have a *period-luminosity relation*. This term means that the longer the period of variation (the interval between successive peaks in brightness), the greater the true average brightness of the star. So an astronomer who measures the apparent magnitude of a Cepheid variable star as it changes over days and weeks, and who thereby determines the period of variability, can readily deduce the true brightness of the star.

Why do we care? Well, knowing the true brightness enables astronomers to determine the distance to the star. After all, the further the star, the dimmer it looks, but it's still the same star with the same true brightness.

Distances dim stars according to the *inverse square law:* When it's twice as far away, it looks four times as faint; when it's three times farther away, it looks nine times as faint; and when it's ten times farther away, it looks 100 times farther away.

Those headlines about the Hubble Space Telescope determining the distance scale and age of the universe come from a Hubble study of Cepheid variable stars. These Cepheids are in faraway galaxies. By tracking their brightness changes and using the period-luminosity relation, the Hubble observers figured out how far away the galaxies are.

RR Lyrae stars

RR Lyrae stars are similar to Cepheids, but not as big and bright. Some of them are located in globular star clusters in our Milky Way, and they have a period-luminosity relation, too.

Globular clusters are huge balls of old stars that were born when the Milky Way was still forming. A few hundred thousand to a million or so stars are all packed in a region of space only 60 to 100 light-years across. Observing the changes in brightness of RR Lyrae stars enables astronomers to estimate their distances, and when the stars are in globular clusters, it tells us how far away the clusters are.

Why is it so important to know the distance of a star cluster? Here goes: All the stars in a single cluster were born from a common cloud at the same time. And they are all at nearly the same distance from Earth, because they are all in the same cluster. So when scientists plot the H-R diagram of stars in a cluster, the diagram is free of errors that could be caused by differences in the distances of the stars. And if we know the distance of the cluster, then all the plotted magnitudes can be converted to actual luminosities, meaning the rates at which the stars are producing energy per second. And those quantities can be directly compared with astrophysical theories of the stars and how they generate their energy. That's the stuff that keeps astrophysicists busy.

Long period variable stars

While astrophysicists celebrate the information gleaned from Cepheid and RR Lyrae variable stars, amateur astronomers delight in observing long period variables, also called Mira stars, or Mira variables. Mira is another name for the star Omicron Ceti, in the constellation Cetus, the Whale, the first known of this type.

Mira variables pulsate like Cepheids, but they have much longer periods, averaging 10 months or more, and the amount by which their visible light changes is even greater. At its brightest, Mira itself can be seen with the naked eye, and at its faintest you need a telescope to spot it. The changes of a long period variable star are also much more variable than those of a Cepheid. The brightest magnitude that a particular star reaches can be quite different from one period to the next. These are easily observed changes that constitute basic scientific information. You can help in these and other variable star studies, as I describe in the last section of this chapter.

Flare stars

Flare stars are little red dwarfs that suffer big explosions, like the flares on the Sun, but more powerful. On the Sun, most flares cannot be seen without the aid of special colored filters because the light of the flare is just a tiny fraction of the total light of the Sun. Only the rare very large "white light" flares can be seen on the Sun without a special filter. (But you still need to use the projection technique or a filter for safe viewing, as I explain in Chapter 10.) But the explosions on flare stars are so bright that the magnitude of the star as a whole changes detectably. Not all red dwarfs have these frequent explosions, but Proxima Centauri, the nearest star beyond our Sun, is a flare star.

Exploding stars: supernovas and cataclysmic variables

The explosions of novas and supernovas are so large that I can't lump them in with the flare stars.

Novas

Novas explode through a build-up process on a white dwarf in a binary system, much like the Type Ia supernovas I describe in the first section of this chapter, but the white dwarf is not destroyed. It just blows its top and then it settles down, sucking more gas off its companion and onto its surface layer. The powerful gravity of the white dwarf compresses and heats this surface layer and after centuries or millennia, off it goes again! At least that's the theory. No one has been around long enough to see an ordinary or *classical* nova explode twice. But similar binary systems exist in which the explosions are not quite as powerful as in a classical nova but recur frequently enough that amateur astronomers are always monitoring them, ready to announce the discovery of a new explosion, and guide professionals to study it. These objects have various names, including *dwarf nova* and *AM Herculis systems*.

Classical novas, dwarf novas, and similar objects are known collectively as *cataclysmic variables.*

There's a nova bright enough to be seen with the naked eye about once a decade, give or take. I studied one in Hercules for my doctoral thesis in 1963. If it hadn't exploded at just the right time, I might still be looking for a thesis topic. Most recently, the bright nova in Vela dazzled astronomers in 1999.

Supernovas

Supernovas throw off nebulae called *supernova remnants,* which fly out in all directions at high speed (see Figure 11-5). The nebula at first consists of the material that once made up the shattered star, less any remaining central object, be it neutron star or black hole. But as it expands into space, it sweeps up interstellar gas, like a snowplow blade that accumulates snow. So after a few thousand years, the supernova remnant is mostly swept-up gas rather than supernova debris.

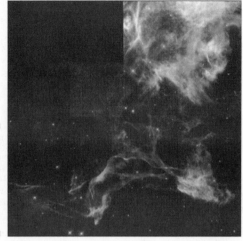

Figure 11-5:
Part of the
Cygnus
Loop, a
supernova
remnant.

Courtesy of NASA

Supernovas are incredibly bright, and rather rare. Astronomers estimate that, in a galaxy like the Milky Way, there is a supernova every 25 to 100 years, but we haven't witnessed a supernova in our home galaxy since Kepler's Star in 1604, before the telescope was invented. Others probably have occurred, but have been obscured from view by dust clouds in the galaxy. A huge southern star called Eta Carinae looks like it may be on the verge of going supernova in the Milky Way, but in astronomers' parlance that means it may explode at any time — within the next million years.

Eclipsing binary stars

Eclipsing binary stars are binary systems whose true brightness doesn't change (unless one of the two stars happens to be a pulsating star, flare star, or other intrinsic variable), but are detected as variable stars by observers on Earth. That's because the *orbital plane* of the system — the plane that contains the orbits of the two stars — is oriented so that it contains our line of sight to the binary system.

If the two stars in a binary have orbital periods of four days, then every four days, the more massive star in the system, usually called "A," passes exactly in front of the other star, as seen from Earth. This blocks all or most of the light from star "B" from reaching us (depending on whether it's larger or smaller than "A" — sometimes the less massive star is larger than its heavy companion), and so the binary looks fainter. That's called a *stellar eclipse.* Two days after this eclipse, star "B" will pass in front of star "A", and we will have another eclipse.

In the section "Born Together, Stay Together: Binary and Multiple Stars," I mention how the orbital velocities are used to learn the masses of the stars.

They can also be used to tell the diameters of stars. Scientists take spectra and learn how fast the stars are orbiting by using the Doppler Effect. And we measure the durations of the eclipses in eclipsing binaries. A stellar eclipse of star "B" begins when the leading edge of "A" starts to pass in front of it. And the eclipse ends when the following edge of "A" finishes passing in front of "B." So the orbital speed times the duration of the eclipse tells you how big star "A" is.

In all these methods, the details are a little more complicated, but it's easy to understand the principles of investigation.

The most famous eclipsing binary is Beta Persei, also known as Algol, the Demon Star.

You won't have a devil of a time observing Algol's eclipses if you're in the Northern Hemisphere — it's a bright star well placed for observation in the northern sky in the fall. You can watch these eclipses without a telescope or even binoculars. Every two days and twenty-one hours, Algol's brightness dims by a little over one magnitude — more than a factor of 2.5 — for about two hours. But you need to know *when* to look for an eclipse. You don't want to stand around in the backyard for almost three days. Check out the fine print in the back pages of *Sky & Telescope*, where they list information for observers. Every so often there's a paragraph called "Minima of Algol," which lists the dates and times when eclipses will occur for a period of a few months.

Minima are the times when variable stars, extrinsic or intrinsic, reach their lowest brightnesses of their current cycles. *Maxima* are the times when they are brightest.

Microlensing events

Sometimes a star that is far from Earth passes precisely in front of another star, which is even farther away. The two stars are unrelated, and may be thousands of light-years from each other. But the gravity of the star in front bends the paths taken by light rays from the star behind in such a way that the distant star appears much brighter to us on Earth for a few days or weeks. This effect is predicted from Einstein's Theory of General Relativity, and it is regularly detected. The effect is called *gravitational lensing,* and when the "lens" or body whose gravity does the light-path bending, is just a star, it is specifically known as *microlensing.* Big-light bending, by the gravity of an entire galaxy or more, is *lensing* without the "micro."

You may think that it's highly unlikely that two unrelated stars would line up perfectly with Earth, and you would be right! Congratulations on that thought. In order to detect such rare events on a regular basis, astronomers use telescopic electronic cameras that can record hundreds of thousands to millions of stars at a time. With that many stars under observation, some foreground star will pass in front of one of them every so often, even though we don't know which.

The trick is to point your telescope at a region where vast numbers of stars can be seen simultaneously in the field of view. Such regions are the Large Magellanic Cloud, a nearby satellite galaxy of the Milky Way, and the central bulge of the Milky Way itself, where there is a whole mess of stars.

Stellar Neighbors to Know

You've already met Proxima Centauri, the nearest star beyond the Sun. It's the third, or outlying member of the Alpha Centauri triple star system.

✔ Alpha Centauri is a bright star in the southern constellation Centaurus (see Figure 11-6). It's a G-type star, a main sequence dwarf about the same color as the Sun but somewhat brighter.

✔ Alpha Centauri's orange companion is a slightly smaller and cooler dwarf named Alpha Centauri B.

✔ The little red dwarf and flare star Proxima is Alpha Centauri C.

The Alpha Centauri system is about 4.4 light-years from Earth, with Proxima on the near side at 4.2 light-years.

Sirius is the brightest star in the night sky. Its official name is Alpha Canis Majoris, in Canis Major, the Great Dog (see Figure 11-7). Slightly south of the Celestial Equator, Sirius is easily visible from most inhabited places on Earth from a distance of 8.5 light-years. It's a white, A-type main sequence star. Sirius is bright enough that folks ask each other, "What's that big star?"

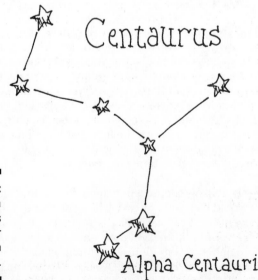

Figure 11-6:
Alpha
Centauri is
in the far
southern
sky.

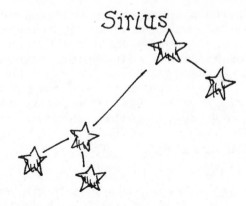

Figure 11-7:
Sirius is top dog in Canis Major.

Like most stars other than the Sun, Sirius has a companion, Sirius B, a white dwarf star. Sirius is known as the Dog Star, and when its tiny companion was discovered, Sirius B was naturally named "the Pup."

There is a legend, and some written records that are open to differing interpretations, that Sirius was seen as a red star a few thousand years ago. Despite much effort, astrophysicists have been unable to explain this color in terms of known physical processes, so naturally we say it wasn't so.

Vega is Alpha Lyrae, the brightest star in Lyra the Lyre. It's high in the sky as seen from temperate northern latitudes (such as the U.S. mainland) on summer evenings, and a star that every self-respecting amateur astronomer knows like the back of his or her hand. Located about 26 light-years from Earth, it's a brilliant, white sparkler and one of the brightest stars in the sky.

Betelgeuse isn't really in the solar neighborhood, it's almost 500 light-years away. But everybody likes its name, which many pronounce "Beetle Juice" (which is as good a way to say it as any), and observers enjoy its deep red color. It's a red supergiant about 50,000 times brighter than the Sun. Although Betelgeuse is formally Alpha Orionis, the brightest star in Orion is Rigel (Beta Orionis).

Helping Scientists by Observing the Stars

Thousands of stars are under watch by astronomers because they vary in brightness or exhibit some other special characteristic. The professional astronomers can't keep up with them all, and that's where you come in. You can monitor some of these stars with your own eyes, with binoculars, or with a telescope.

You need to be able to recognize the stars and to judge their brightnesses. The brightness of many stars changes so significantly — by a factor of two, or ten, or even by hundreds — that eye estimates are sufficiently accurate to keep track of them. The trick is to use a *comparison chart,* a map of the sky that shows the position of the variable star and the positions and magnitudes of *comparison stars.* Comparison stars are stars that don't vary in brightness, and whose brightnesses have already been measured.

The American Association of Variable Star Observers (AAVSO) offers a wealth of information explaining how to observe variable stars. Their Web site is www.aavso.org. They offer help to observation novices. They sell an inexpensive star atlas and hundreds of individual comparison charts for different variable stars. You can buy charts from them for just 25 cents each, or you can download them from the AAVSO Web site for free.

The AAVSO administers a Nova Search and a Supernova Search that you can join as you grow skilled in celestial observations.

✔ **Nova Search:** This program requires only patience, care, and a pair of binoculars. When you join, they will assign you a small part of the sky. Then, as often you can on clear nights, check your sky section. Scan it slowly with binoculars as you check the patterns of faint stars against those on your star charts.

If you find a "new star" (the original meaning of *nova* from the Latin) that's not on your charts, report it as fast as possible, preferably by e-mail. You may have discovered a real *nova,* an explosion in a particular kind of binary star. But you may want to wait a few hours to see if the "nova" moves. If it moves slightly with respect to other stars in the field of view, it's not a star at all. It's probably an asteroid or a faint comet. And other mistakes can occur. Back in the early 1950s, my pal Charlie and I sent a telegram to the AAVSO, announcing our discovery of a nova that we viewed through a telescope from a rooftop in Brooklyn. We thought that it was a nova because it wasn't moving and it wasn't on the chart. But fame and fortune evaded us: It was a star that had been inadvertently left off the chart.

✔ **Supernova Search:** This program is for advanced amateurs. After a few years you may be ready for it. You need a pretty good telescope for this kind of observation. And preferably, you'll have an electronic camera to make photographs through the telescope. Instead of monitoring for nova explosions a patch of sky in our own Milky Way galaxy, you'll look at distant galaxies one at a time, searching for a bright spot that may suddenly appear where you saw none the last time you looked. The bright spot is a *supernova.* Because a supernova is so much brighter than a nova, a supernova can be easily seen, although in a distant galaxy.

Chapter 12

Galaxies: The Milky Way and Beyond

. . . .

In This Chapter

▶ Tasting the Milky Way, its star clusters, and its nebulae

▶ Classifying galaxies by shape and size

▶ Sorting galaxies into groups and clusters

▶ Contemplating superclusters, Great Walls, and cosmic voids

. . . .

*O*ur solar system is a tiny part of the Milky Way galaxy, a great system of hundreds of billions of stars, thousands of nebulae, and hundreds of star clusters. The Milky Way, in turn, is one of the largest components of the Local Group of Galaxies. Beyond the Local Group is the Virgo Cluster, the nearest large cluster of galaxies — a full 50 million light-years from Earth. As scientists look out into the universe to much greater distances, they see superclusters, immense systems that contain many individual clusters of galaxies. So far, no superclusters of superclusters have been found, but *Great Walls,* which are immensely long superclusters, exist. And much of the universe seems to be taken up by cosmic voids, which contain few detectable galaxies.

This chapter introduces you to the Milky Way and its most important parts, and takes you systematically farther into the universe to meet other types of galaxies and see how they are organized in space.

Unwrapping the Milky Way: Earth's Galactic Home

Meet the Milky Way! It's a lot bigger than the candy bar, if not as sweet. But it does have a creamy-looking center. The Milky Way is that wide band of diffuse light that you can see best on clear summer and winter nights.

A stream of milk through the universe was as good as any explanation for the Milky Way until 1610, when Galileo scrutinized it with a telescope. He found that the Milky Way was nothing to lap about: It consists of an immense number of dim stars that blend together into one large fuzzy cloud as visible to the eye. The telescope was a definite improvement.

As I explain later in this chapter, galaxies are the basic building blocks of the universe, and the Milky Way is a good-sized block. It contains almost everything in the universe that's visible to the naked eye and lots that isn't — from Earth and its solar system to the stars of the solar neighborhood, the visible stars in the constellations, and all the stars that blend together to make that milky stream in the night sky. In addition, the Milky Way contains nearly every single nebula that you can see without telescopic aid, and plenty that you can't see.

The Milky Way is a big galaxy! Besides loose stars, it contains hundreds of star clusters, such as the Pleiades and Hyades in the constellation Taurus, and, for lucky viewers in Australia, South America, and other points far south, the Jewel Box in Crux, the Southern Cross, and the great round Omega Centauri.

What shape is the Milky Way?

The Milky Way is a spiral galaxy, consisting of a pizza-shaped formation of billions of stars (the *galactic disk*) that contains the spiral arms. The arms are shaped roughly like the streams of water from a rotating lawn sprinkler and contain lots of bright, young, blue and white stars and gas clouds. Groups of young, hot stars (called *associations*) dot the spiral arms in the galactic disk like pepperoni slices on a pizza. Bright and dark nebulae seem to mushroom in the arms as well. Between the arms are the *interarm regions* (not all astronomical terms are as catchy as Barnacle Bill, a rock on Mars, or the Red Rectangle, a nebula that's actually shaped like an hourglass).

The now murky Milky Way

Everyone used to recognize the Milky Way, but now many people don't see it or don't know it because they live in or near cities, and the bright lights obscure it.

The solution to seeing the Milky Way without light pollution is to get out to the mountains or the shore on your vacation and check out a darker sky than you can see at home. The light of the full Moon interferes with seeing the Milky Way, too, so plan your holiday around the time of the new Moon, when there is little or no moonlight. The Milky Way is most prominent in the sky during summer and winter and least visible in spring and fall.

Seeing beyond the Milky Way

The three objects readily visible with the naked eye that are beyond the Milky Way are the Large and Small Magellanic Clouds (two nearby galaxies that can be seen from the Southern Hemisphere), and the Andromeda Galaxy. Some people blessed with excellent vision (and many others who are just trying to impress their friends) say that they can see the Triangulum Galaxy as well. Both the Andromeda and Triangulum Galaxies are about two million light-years from Earth, but Andromeda is bigger and brighter.

I'm counting the Large Magellanic Cloud as one object, but in fact it contains a huge bright nebula, the Tarantula, which can also be distinguished by the eye. For a few months in 1997, a bright supernova in the Large Cloud, Supernova 1987A, could be seen there, too.

At the center of our galaxy is a place called (you guessed it!) the *galactic center*. Centered on the center is the galactic bulge, which would put a sumo wrestler to shame. The *galactic bulge* is a roughly spherical formation of millions of mostly orange and red stars, sitting like a great meatball at the center of the galactic disk and extending far above and below it. And at the center is Sagittarius A*, a supermassive black hole. Figure 12-1 presents a model of the Milky Way with its ingredients.

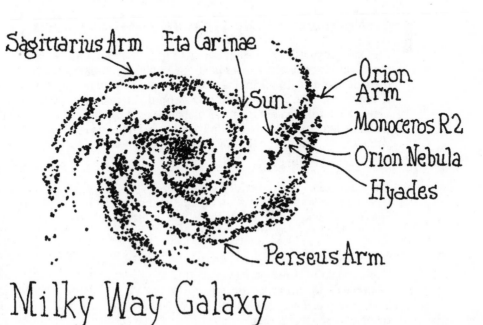

Figure 12-1:
The Milky Way is a spiral galaxy with arms encircling the galactic center.

The flat imaginary surface or mid-plane of the galactic disk is called the *galactic plane,* and the circle that represents its intersection with the sky, as visible from Earth, is called the *galactic equator.*

Sometimes the position of an object is listed in galactic coordinates, instead of right ascension and declination (coordinates that I define in Chapter 1). The galactic coordinates are *Galactic Latitude,* which is measured in degrees north or south of the galactic equator, and *Galactic Longitude,* which is measured in degrees along the galactic equator.

Galactic Longitude starts at the direction to the galactic center, which is zero degrees Longitude. (Actually, the zero point of Galactic Longitude is slightly off the galactic center, because it's the place where the galactic center was believed to be in 1959, and we know better now.) Galactic Longitude proceeds along the galactic equator from the constellation Sagittarius into Aquila, Cygnus, and Cassiopeia; it then goes on through Auriga, Canis Major, Carina, and Centaurus — all the way to 360 degrees Longitude, back at the galactic center. When you look with your binoculars at the constellations I just named, you'll see more stars, star clusters, and nebulae than elsewhere in the sky.

The "plane truth" is that the constellations that the galactic plane intersects are among the finest sights in the sky.

Where can you find the Milky Way?

The Milky Way isn't a certain distance from the Sun and Earth; it contains them. But the galactic center is about 25,000 light-years from Earth. Recent measurements with a radio telescope called the *Very Long Baseline Array* revealed that the solar system takes about 226 million years to make one orbit around the galactic center. This information cleared up a big discrepancy: Until then, scientists didn't know whether this interval, called the *galactic year,* was 200 million years or 250 million years. Now, astronomers can set the calendars straight.

The outskirts of the galaxy, or *galactic rim* — as science fiction fans know it — is, at its closest part, about an equal distance in the opposite direction from Sagittarius. The disk of the Milky Way is pretty much identical with that milky band of light in the sky.

The Milky Way is about 169,000 light-years from the Large Magellanic Cloud, about 2 million light-years from Andromeda, and about 50 million light-years from the nearest big cluster of galaxies, the Virgo Cluster. It's also smack dab in the middle of a little cluster of galaxies (sizes are relative here), the Local Group, all of which I discuss later in this chapter.

How and when was the Milky Way formed?

The Milky Way is probably almost as old as the universe, and certainly more than the 12 billion years that is the estimated age of some of its oldest stars. Some estimated ages are even older. No one we know was around at the time — Earth didn't even exist yet — so these estimates are very rough.

The Milky Way is the shape it is and the size it is because, in the universe, gravity rules. Long ago, gravity caused a gigantic cloud of primordial gas to fall together and condense. As little clumps inside the collapsing cloud collapsed even faster than the cloud as a whole, stars were produced. Because the big cloud must have started out turning very slowly, it would have rotated ever faster as it became smaller, flattening much of it into the present day spiral disk formation. And before you knew it, *voila, la voie lactee* (that's French for "there it is, the Milky Way").

If you have a better theory, become an astronomer yourself and you can write your own book someday — in science, theories make the world go 'round and maybe even the galaxy.

Galactic Associates: Star Clusters

Star clusters are just bunches of stars, located in and around a galaxy. They are not chance associations (even though one type of star cluster is called an "association"), but groups of stars that were born together from a common cloud and that, in most cases, are held together by gravity.

The three main types of star clusters are open clusters, globular clusters, and OB associations.

For superb images of star clusters, consult the Anglo-Australian Observatory at www.aao.gov.au/local/www/dfm/cluster_frames.html, or treat yourself to a "coffee-table" book — *The Invisible Universe* by David Malin (Bulfinch Press, 1999) — that contains a collection of the greatest photographs from that observatory.

Open clusters

Open clusters contain dozens to thousands of stars, have no particular shape, and are located in the disk of the Milky Way. Typical open clusters are about 30 light-years across. They are not highly concentrated (if at all) toward their centers, unlike globular clusters, and typically are much younger. They are great targets for viewing with small telescopes and binoculars, and some can be seen with the naked eye.

The most famous and easily seen open clusters in the northern sky are

✔ The Pleiades, in the northwest corner of Taurus, the Bull.

The Pleiades, also known as the Seven Sisters, looks with the naked eye like a tiny dipper. You can compare your eyesight with a friend's by seeing how many stars you can count in the Pleiades, which is M45, the 45th object in the *Messier Catalog* (see Chapter 1). Then gaze through binoculars and see how many more you can find. The brightest star in the Pleiades is Eta Tauri (3rd magnitude), also called Alcyone. (See Chapter 1 if you need an explanation of magnitude.)

✔ The Hyades, also in Taurus.

The Hyades, also a great sight for the naked eye, includes most of the stars that make up the "V" shape in the head of Taurus. You can't miss it, because the V includes the bright red giant Aldebaran (1st magnitude), which is Alpha Tauri (see Figure 12-2). Aldebaran actually is not in the Hyades, it's far beyond it, but in the same direction as seen from Earth.

The Hyades looks much bigger than the Pleiades because it's only about 150 light-years from Earth, versus about 400 light-years for the Pleiades.

✔ The Double Cluster, in Perseus, the Hero.

The Double Cluster is a beautiful sight in binoculars and especially in a small telescope. Its two clusters are NGC 869 and 884, each probably over 7,000 light-years from Earth. NGC stands for *New General Catalogue,* which was new when it first appeared in 1888.

✔ The Beehive, in Cancer, the Crab.

The Beehive (Messier 44) is the main attraction of Cancer, a constellation composed of dim stars. It looks like a nice fuzzy patch with the naked eye, and a swarm of many stars through binoculars.

For viewers in the Southern Hemisphere, the finest open clusters include

✔ NGC 6231, in Scorpius, the Scorpion.

NGC 6231 is a southern object, but it can be readily seen from much of the United States in the evening during the summer. You need to be in a dark place, with an unobstructed southern horizon.

✔ The Jewel Box, in Crux, the Southern Cross.

The Jewel Box includes the bright star Kappa Crucis. The Southern Cross is a perennial favorite of viewers in the Southern Hemisphere. If you take a cruise through the South Seas, insist that the ship have an astronomy lecturer on board. He or she will gladly point out the Southern Cross; with binoculars, you can enjoy the fine sight of the Jewel Box.

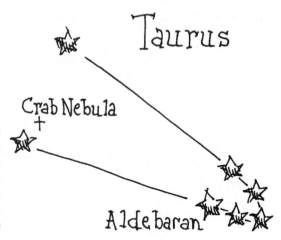

Figure 12-2:
Taurus
contains the
red giant
Aldebaran.

Globular clusters

Globular clusters are the retirement homes of the Milky Way galaxy. They are
about as ancient as the galaxy itself (some experts think that they were the
first objects to form in the Milky Way), so they are made of very old stars,
including many red giants and white dwarfs (see Chapter 11). The stars you
can see in a globular cluster with your telescope are mostly red giants. With
bigger telescopes, orange and red main sequence dwarf stars are readily
observed. Only the Hubble Space Telescope and other very powerful instruments can pick out more than a handful of the much fainter white dwarfs.

A typical globular star cluster contains from a hundred thousand to a million
or more stars, all packed in a ball (hence the term "globular") of just 60 to
100 light-years in diameter. And, the closer to the center, the more tightly the
stars are packed (see Figure 12-3). This high degree of concentration and the
great number of stars distinguish a globular cluster from an open cluster.

Figure 12-3:
Globular
cluster G1
in the
Andromeda
Galaxy.

Courtesy of NASA

Another key difference is that open clusters are distributed across the galactic disk in a great flat pattern, but globular clusters are arranged spherically around the center of the Milky Way, with many high above and deep below the galactic plane. These clusters, too, are concentrated toward the galactic center, but many of the globulars that are easiest to see are well above or below the galactic plane.

The best globular clusters for viewing in the northern sky are

- ✔ Messier 13, in Hercules, representing the mythical character of the same name
- ✔ Messier 15, in Pegasus, the Winged Horse

You can spot both M13 and M15 with the naked eye under suitably dark sky conditions, but you'll need to reassure yourself with binoculars or a small telescope, which show them as fuzzy spots, larger than stars. Use a star chart (such as *Norton's Star Atlas* by Arthur P. Norton; 19th Edition edited by Ian Ridpath and published by Longman Publishing Group, 1998) to locate them.

Observers in the Northern Hemisphere have been cheated out of the best globular star clusters, because by far the two biggest and brightest are in the deep southern sky:

- ✔ Omega Centauri, in Centaurus, the Centaur
- ✔ 47 Tucanae, in Tucana, the Toucan

They are spectacular sights when viewed through small binoculars, and just about worth the trip to South America, South Africa, Australia, or other places where they can be readily seen.

OB associations

OB associations are loose groupings with dozens of stars of spectral types O and B and sometimes fainter, cooler stars (see Chapter 11 for more about spectral types). Unlike open clusters and globular clusters, gravity doesn't hold these associations together; with time, the stars in them move away from each other, and the association is dissolved, like a limited partnership that has reached its limit. OB associations are located close to the galactic plane.

Many of the bright young stars in the constellation Orion (which is just southwest of the galactic plane) are members of the Orion OB association.

Nebulae: Bright Shining Clouds and Some That Don't

A nebula is a cloud of gas and dust in space. (*Dust* means microscopic solid particles, which may be made of silicate rock, carbon, ice, or various combinations of those substances.) The plural of nebula is nebulas or nebulae. As I note in Chapter 11, some nebulae play an important role in star formation; others are produced by stars that are on their deathbeds. Between the cradle and the grave, nebulae come in a number of varieties.

I describe the most important nebulae in the following list:

- ✔ **H II regions** are nebulae in which the hydrogen is ionized, meaning that the hydrogen has lost its electron. (A hydrogen atom has one proton and one electron.) The gas in an H II region is hot, ionized, and glowing due to the effects of ultraviolet radiation from nearby O or B stars. All of the large bright nebulae that you can see through binoculars are H II regions. *H II* is a term that refers to the spectrum of ionized hydrogen.

- ✔ **Dark nebulae** are the dust bunnies of the Milky Way. They are clouds of gas and dust that are not shining. Their hydrogen is neutral, meaning that it has not lost its electron. The term H I region refers to a nebula in which the hydrogen is neutral — another name for these dark nebulae.

- ✔ **Reflection nebulae** are composed of dust and cool, neutral hydrogen. They shine by the reflected light of nearby stars. If those stars were not nearby, these objects would be dark nebulae.

- ✔ **Giant molecular clouds** are the largest objects in the Milky Way. But they are cold and dark, and we wouldn't even know that they were there if it weren't for the data gathered by radio telescopes, which can detect emissions of faint radio waves from molecules such as carbon monoxide (CO). Like all of the other nebulae, giant molecular clouds are mostly made of hydrogen, but are often studied by means of their trace components such as CO. The hydrogen in these giant clouds is molecular, with the designation H_2, which means that each molecule consists of two neutral hydrogen atoms.

One of the most exciting discoveries in nebular studies in recent decades was that bright H II regions, such as the Orion Nebula, are just little hot places on the peripheries of giant molecular clouds. For centuries, people could see the Orion Nebula, and had no idea that it was no more than a bright pimple on a huge invisible object, the Orion Molecular Cloud. But now we know. New stars are born in molecular clouds, and when they get hot enough, they ionize their immediate surroundings, turning them into H II regions. Where the dust in a molecular cloud is thick enough to cut off the light of many or most of the stars behind the cloud, as visible from Earth, we call that part of the molecular cloud a dark nebula.

> ✔ **Planetary nebulae,** as I note in Chapter 11, are the atmospheres of old stars that started out resembling the Sun but ditched their outer layers as they entered their death throes. I talk more about planetary nebulae in the next section.
>
> ✔ **Supernova remnants** are nebulae that begin as material ejected from massive stellar explosions, as I describe in Chapter 11. I cover supernovas in more detail later in this chapter.

H II regions, dark nebulae, giant molecular clouds, and many of the reflection nebulae are located in or near the Milky Way's galactic disk.

Planetary nebulae

Planetary nebulae are the atmospheres of old stars that started out resembling the Sun but then expelled their outer atmospheric layers. The nebulae are ionized and made to glow by ultraviolet light from the hot little stars at their centers, which are all that remain of the former suns. These nebulae are expanding into space and fading as they grow larger.

For decades, astronomers believed that many or most planetary nebulae were spherical or roughly so. But now it's known that most are bipolar, meaning that they consist of two round lobes projecting from opposite sides of the central star. Those planetary nebulae that look spherical, such as the Ring Nebula in the constellation Lyra (see Figure 12-4), are bipolar, too, but the axis down the center of the lobes happens to point at Earth, and so, like a dumbbell viewed end-on, it looks circular. Astronomers took many years to figure this out, so maybe we were dumbbells, too. Planetary nebulae can be well off the galactic plane, unlike H II regions.

A galactic goof-up

In earlier times, right up through the 1950s, the term "nebula" was also used to refer to a galaxy, because until the 1920s, galaxies beyond the Milky Way were thought to be nebulae in the Milky Way. Astronomers believed in the existence of only one galaxy, that is, Earth's galaxy: the Milky Way.

It took a few dozen years for the change in understanding to prevail in the language of astronomy, so the authors of astronomy books have only recently stopped referring to the Andromeda Galaxy as the Andromeda Nebula.

Edwin P. Hubble, for whom the telescope and many other things in astronomy are named, wrote a famous book, *The Realm of the Nebulae*. It was all about galaxies, not nebulae as we use the term now. Among his many achievements, Hubble proved that the Andromeda Nebula was a galaxy full of stars, not a big cloud of gas. A former boxer, he fought in World War I, smoked a pipe, and was said to bully some of the other astronomers at Mount Wilson Observatory, but his discoveries were no bull.

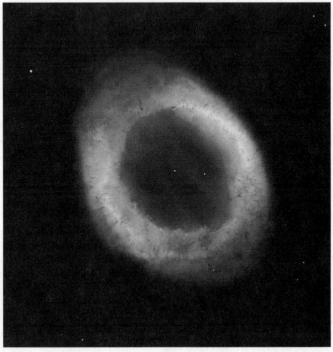

Figure 12-4:
The Ring
Nebula in
Lyra.

Curious point: Respectively related and unrelated to planetary nebulae are *protoplanetary nebulae*, which are much studied by astrophysicists. One type of protoplanetary nebula is the early stage of a planetary nebula, that is, a phase in the death of a star. The other type of protoplanetary nebula is the birth cloud of a solar system of a star and its planets. It's pretty silly of astronomers to use the same term to refer to two completely different kinds of objects, but nobody's perfect. We may need another Edwin P. Hubble to bully us into some sensible nomenclature.

Supernova remnants

Supernova remnants begin as material ejected from massive stellar explosions. A young supernova remnant is composed almost exclusively of the shattered remains of the exploded star that produced it. But as the gas moves outward through interstellar space, it resembles a rolling stone that *does* gather moss. A snowplow effect is created as the expanding supernova remnant pushes along and accumulates the thin gas of interstellar space. By the time the supernova remnant becomes old — tens of thousands of years later — the nebula is overwhelmingly composed of this "plowed up" interstellar gas, and the remains of the exploded star are mere traces.

Supernova remnants are found along or near the galactic plane of the Milky Way.

Nebulae worth watching

The following are some of the best, brightest (or for dark nebulae, darkest), and most beautiful nebulae visible from northern latitudes, including some objects that are not too far south of the celestial equator:

- ✔ The Orion Nebula, Messier 42, in Orion, the Hunter.

  An H II region, the Orion Nebula is very easily seen with the naked eye as a fuzzy spot in the sword of Orion. The nebula looks fine in binoculars and spectacular in a small telescope. The telescope also shows the Trapezium, a bright quadruple star (see Chapter 11) in the nebula.

- ✔ The Ring Nebula, Messier 57, in Lyra.

  The Ring Nebula is a planetary nebula high in the sky at northern temperate latitudes on a summer evening. Like all planetary nebulae, you need to use a star chart to find it with your telescope, unless you have a computer assisted telescope such as the Meade ETX-90/EC (see Chapter 3), which will point right to the nebula when so commanded.

- ✔ The Dumbbell Nebula, Messier 27, in Vulpecula, the Little Fox.

  The Dumbbell Nebula, along with the Ring Nebula, is among the easiest planetary nebulae to spot with a small telescope. It's well placed for observation in the summer and fall.

- ✔ The Crab Nebula, Messier 1, in Taurus.

  The Crab Nebula is the remains of a supernova that exploded in the year 1054, as seen from Earth. The Crab Nebula is another fuzzy spot if viewed with a small telescope, but a big telescope shows two stars near its center. One star is not associated with the Crab; it's just located along the same line of sight. The other star is the pulsar that remains from the supernova explosion. It's spinning 30 times per second, and one or the other of its two lighthouse beams sweeps across Earth every $\frac{1}{60}$ of a second, with the same frequency as your 60-cycle, alternating-current household electricity. (I think that's a "powerful" analogy.)

- ✔ The North American Nebula, NGC 7000, in Cygnus, the Swan.

  The North American Nebula is a faint but large H II region, which can be seen with the naked eye of an attentive viewer on a dark summer evening. For ease in detecting it, use averted vision — look out of the corner of your eye. Its name comes from its shape.

- ✔ The Northern Coal Sack, in Cygnus, the Swan.

  The Northern Coal Sack is a dark nebula near Deneb, which is Alpha Cygni, the brightest star in Cygnus. You can identify it by eye as a dark blotch against the brighter background of the Milky Way.

At moderate southern declinations, but visible from most places in both Northern and Southern Hemispheres and not to be missed are

- ✓ The Lagoon Nebula, Messier 8, in Sagittarius, the Archer.
- ✓ The Trifid Nebula, Messier 20, in Sagittarius.

  The Lagoon Nebula and the Trifid Nebula are large, bright H II regions that can be seen in the same field of view of your binoculars. They are well placed for observation on summer evenings. A color photo shows that the Trifid has a bright red region and a separate, fainter blue region. The red area is the H II region and the blue zone is a reflection nebula.

Great nebulae of the deep Southern Hemisphere include

- ✓ The Tarantula Nebula, in Dorado, the Goldfish.

  The Tarantula Nebula is not in the Milky Way at all, but in the Large Magellanic Cloud galaxy. But it is such a huge and brilliant H II region that it's conspicuous to the naked eye for viewers in temperate and far southern latitudes. The Tarantula is another object to observe if you take a South Seas cruise. Trust me; you won't carp at Dorado.

- ✓ The Carina Nebula, in Carina, the Ship's Keel.

  The Carina Nebula, in the region of the huge, unstable star Eta Carinae (see Chapter 11), is a large, bright H II region.

- ✓ The Coal Sack, in Crux.

  The Coal Sack, a dark nebula, is a large black patch, several degrees on a side, in the Milky Way in Crux. You can't miss it on a clear night with a dark sky, as long as you are deep in the Southern Hemisphere.

- ✓ The Eight-Burst Nebula, NGC 3132, in Vela, the Sail.

  The Eight-Burst Nebula is a planetary nebula in the far southern sky.

Galaxies: Islands in the Universe

A large galaxy consists of thousands of star clusters and billions to trillions of individual stars, all held together by gravity. The Milky Way fits this bill; it's a large spiral system. But galaxies come in many other shapes and sizes (see Figure 12-5).

The main types of galaxies, based on shape and size, are

- ✓ Spiral galaxies
- ✓ Barred spiral galaxies
- ✓ Lenticular galaxies

✔ Elliptical galaxies

✔ Irregular galaxies

✔ Dwarf galaxies

✔ Low surface brightness galaxies

Spiral, barred, and lenticular galaxies

Spiral galaxies are disk-shaped, with spiral arms winding through their disks. They resemble the Milky Way, but the arms may be wound more tightly than our galaxy's spiral arms, or less so. And the central bulge of stars in another spiral galaxy may be more prominent or less prominent, compared to the spiral arms.

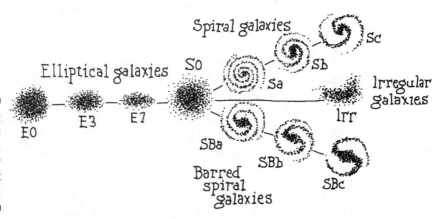

Figure 12-5: Galaxies come in many different types.

Shape is one of the earliest ways by which galaxies have been classified. In fact, one of the most used classification systems, for many years, was called the *Hubble types,* devised by you know whom. (Edwin P. Hubble, 1889–1953, made a number of significant contributions to the science of astronomy in the 20th century, leading NASA to name the Hubble Space Telescope for him.)

Spiral galaxies are distinguished by having lots of interstellar gas, nebulae, OB associations, and open clusters, in addition to globular clusters.

Barred spiral galaxies are spiral galaxies in which the spiral arms don't seem to emerge from the galaxy center but from the ends of a linear or football-shaped cloud of stars that straddles the center. This star cloud is called the *bar.* Gas from outer parts of the galaxy may be funneled toward the center through the bar. This process can form new stars that make the galaxy's central bulge, well, bulge even more.

Lenticular (lens-shaped) galaxies are flattened systems, with galactic disks, just like spiral galaxies. They contain gas and dust. But they don't have spiral arms.

Elliptical galaxies

Eiliptical galaxies are football-shaped, and that definition includes both U.S. football and soccer. Some ellipticals, in other words, are ellipsoidal in shape, roughly like a U.S. football, and some are spherical, like a soccer ball. They can be beautiful sights, and I get a kick out of them. They contain lots of old stars and globular star clusters, but not much else.

Elliptical galaxies are systems in which star formation has largely or totally ceased. There are no H II regions, young star clusters, or OB associations. Imagine living in one of these dull galaxies, with nothing like the Orion Nebula to entertain you or give birth to new stars. And probably not much on TV, either.

The production of new stars may have ended in an elliptical galaxy because all the gas has been used up in making the stars that are already there. Or it may have ended because something blew out or swept out all the remaining gas suitable for making more stars. I put it that way because some elliptical galaxies, although they show no H II regions or groups of young stars overall, do have some extremely hot gas, so thin and hot that it shines only in X rays. Gas like that won't readily condense into stars. And, to tell the truth, some elliptical galaxies do display a number of bluish star clusters, which appear to be very young globular star clusters, much younger than any in the Milky Way.

A leading theory of elliptical galaxies, or at least of *some* elliptical galaxies, is that they have been produced by the collision and merger of smaller galaxies. The collision of two spiral galaxies, for example, could produce a large elliptical galaxy, and shock waves from the event might compress large molecular clouds in the spirals, giving birth to huge clusters of hot, young stars, perhaps the very bluish star clusters found in some ellipticals. But the collision of a small galaxy with a big spiral might just lead to the latter swallowing the former. Then the spiral's central bulge would get even bigger.

As astronomers look out into space, we can see many examples of colliding and merging galaxies, and the further back we look, the more prevalent they seem to be. Apparently, galaxy collisions were common in the early universe, and may have helped to shape many of the galaxies that we see today.

A galaxy is a galaxy is a galaxy

Writing "galaxy" and "galaxies" over and over is repetitive. But there just isn't any synonym for "galaxy." Some uninformed folks (or their editors) write "star cluster" to try and vary their prose, but that's plain wrong. And a large group of galaxies is not a "galactic cluster," which is a term meaning an open cluster of stars inside a galaxy. That large group is a *cluster of galaxies*, sometimes written, "galaxy cluster." It's composed of galaxies and thus *galaxian*, but it isn't galactic.

Irregular, dwarf, and low surface brightness galaxies

Irregular galaxies have shapes that tend to be, well, strictly irregular. You may find the glimmerings of a little spiral structure in one of them, or you may not. They generally have lots of interstellar gas, with new stars forming all the time. And they are usually smaller than full-sized spirals and ellipticals, with many fewer stars.

Dwarf galaxies are just what the name implies, itty bitty little galaxies that may be mere thousands of light-years across, or even less. The types of dwarf galaxies include dwarf ellipticals, dwarf spheroidals, dwarf irregulars, and apparently also (although this subject has been a bit controversial) dwarf spirals.

In our immediate neck of the woods, the Local Group of Galaxies (more about it in the next section), the most common galaxies are the dwarf galaxies — just as in the Milky Way, the most common stars are the smallest stars, the red dwarfs. The same is probably true far off in space, but it's hard to tell, because the dwarf galaxies are much harder to see and count at great distances than are the full-sized galaxies.

Low surface brightness galaxies were recognized as a major class of object during the 1990s. They can be as large as most other galaxies, but barely shine at all. Although they are full of gas, they have not produced many stars, so they don't shine very brightly. They were missed for decades in surveys of the sky, and astronomers are only beginning to pick them up now with advanced electronic cameras. They aren't suitable targets for home telescopes, but I thought you should know about them. Who knows what else is out there that we haven't spotted yet?

Some astrophysicists think that much of the mass in the universe might be present in the form of low surface brightness galaxies that just haven't been counted properly, like groups of people who are said to be under-represented in the U.S. Census.

Great galaxies for gawkers

The best galaxies for viewing from the Northern Hemisphere include:

- The Andromeda Galaxy (Messier 31), in Andromeda, a constellation named for an Ethiopian princess in Greek mythology (Figure 12-6 shows this galaxy).

  The Andromeda Galaxy is also called the Great Spiral Galaxy in Andromeda, and was long known as the Great Spiral Nebula in Andromeda. It's another fuzzy patch to the naked eye; it appears in the northern autumn evening sky. But from a dark observing site, you can trace it across about three degrees on the sky with your binoculars, or about six times the width of the full Moon. Don't try to view it during the full Moon, wait until the moon is barely illuminated or, better yet, below the horizon. The darker the night, the more of the Andromeda Galaxy you can see.

- NGC 205 and Messier 32, in Andromeda.

  NGC 205 and Messier 32 are elliptical galaxies that are small, close companions of the Andromeda Galaxy. Some experts call them both dwarf elliptical galaxies and some don't. (I wish they would make up their minds.) M32 is spheroidal in shape, and NGC 205 is ellipsoidal.

- The Triangulum, or Pinwheel Galaxy (Messier 33), in Triangulum, the Triangle.

  The Triangulum, or Pinwheel Galaxy, is another large, bright, nearby spiral galaxy, smaller and a little dimmer than the Andromeda Galaxy and also a fine sight through binoculars in the fall.

- The Whirlpool Galaxy (Messier 51), in Canes Venatici, the Venetian Hunting Dogs.

  The Whirlpool Galaxy is further away and fainter than the Andromeda and Triangulum Galaxies, but is a more glorious sight in a high quality small telescope. It is a *face-on* spiral, meaning that the galactic disk of the Whirlpool is pretty much at right angles to our line of sight from Earth; we are looking right down (or up) on it. With the larger telescopes at a star party, you should be able to make out its spiral structure from a distance of about 15 million light-years. Messier 51 was where scientists discovered the spiral structure of galaxies, long before we knew those "nebulae" were galaxies. Look for it on an evening in the spring.

- The Sombrero Galaxy (Messier 94), in Virgo, the Virgin.

  The Sombrero Galaxy is a bright, edge-on spiral galaxy. It's "brim" is the Sombrero's galactic disk and a dark stripe appears on the brim because we are looking right at the band of dark nebulae or coal sacks in that galactic disk. Look for it in the spring, too; it's almost three times farther away than the Whirlpool, but still a good sight in a telescope.

M 31

The Andromeda Galaxy

Figure 12-6:
The
Andromeda
Galaxy

| 40,000 LY | 2,000 LY | 40 LIGHT–YEARS |
| Ground View of Galaxy | Ground View of Galaxy Core | HST View of Galaxy Nucleus |

Courtesy of NASA

The following lists the finest galaxies for observers in the Southern Hemisphere:

✔ The Large and Small Magellanic Clouds (LMC and SMC) are irregular galaxies that are satellites of the Milky Way. The Large Cloud is not only larger, it is also closer to Earth. It's a mere 169,000 light-years (give or take) from us. In fact, for many years the LMC was believed to be the closest galaxy beyond the Milky Way. (Today, scientists know that a dim, miserable rendition of a galaxy, called the Sagittarius Dwarf Galaxy, is even closer, but it's barely discernible in telescopic photos, because it's being absorbed into the Milky Way. So long, Sagittarius, we hardly knew ye!)

The LMC and SMC actually look like clouds in the night sky. They are that big and bright, and are circumpolar in much of the Southern Hemisphere. In other words, at far southern latitudes, they never set below the horizon. If you get far enough south in South America or other places in the Southern Hemisphere, the Clouds are visible on any clear night of the year. Sweep through them with binoculars and see how many star clusters and nebulae you think that you can see.

✔ The Sculpter Galaxy (NGC 253) is a large, bright spiral galaxy.

✔ Centaurus A (NGC 5128) is a huge galaxy that has a peculiar appearance. It's spheroidal, but has a thick band of dark dust across its middle. It's a powerful source of radio waves as received with radio telescopes. Theorists have gone back and forth as to whether or not it's an example of colliding galaxies. I think that it has probably swallowed a smaller galaxy or two in its time, so watch from a safe distance.

The Local Group of Galaxies

The Local Group of Galaxies, called Local Group for short, consists of two large spirals (the Milky Way and the Andromeda Galaxy), a smaller spiral (the Triangulum Galaxy), their satellites (including the Large and Small Magellanic Clouds, as well as M32 and NGC 205), and about two dozen dwarf galaxies.

The Local Group isn't much as assemblages of galaxies go, but it's home. It is the largest structure that we on Earth are gravitationally bound to. That means that Earth is not flying away from the Local Group as the universe expands. Just as the solar system isn't getting any bigger — because the Sun's gravity prevents the planets from moving outward or escaping — the Local Group is held together by the gravity of the three spiral galaxies and the smaller members. But all other groups and clusters of galaxies and distant individual galaxies throughout the universe outside the Local Group's gravitational pull *are* moving away from the Local Group at rates determined by a formula called Hubble's Law (named for the astronomer, not the telescope). Chapter 16 explains more about the movement away.

The Local Group is about one megaparsec across and centered near the Milky Way. A *parsec* is a dimension in space equal to 3.26 light-years. And *mega* means million. So the Local Group is about 3.26 million light-years, or about 19 trillion miles wide. That dimension may sound large, but it's minuscule compared to the observable extent of the universe beyond.

Clusters and superclusters of galaxies are much larger than the Local Group, and can be easily spotted across billions of light-years in space. But the majority of all the galaxies in the universe, at least the readily visible ones, are located in small groups with only dozens of members or less, like the Local Group (which has about 30). So we appear to be in an average condition, as galaxy neighborhoods go.

Clusters of galaxies

Most galaxies may be in small groups like The Local Group, but as astronomers survey the distant heavens with professional observatory telescopes, the formations that stand out are the clusters of galaxies. Most prominent are the so-called *rich clusters,* with hundreds and even thousands of galaxy members, each with its own complement of billions of stars.

The nearest large cluster of galaxies is the Virgo Cluster, spread out across the constellation of the same name and adjacent constellations. The cluster is about 50 million light-years away and contains hundreds of known galaxies.

You can observe some of the biggest and brightest member galaxies of the Virgo Cluster with your own telescope. Messier 87 is one of the best sights. It's a spheroidal-shaped giant elliptical galaxy with a powerful jet of matter flying out at its center from the vicinity of a supermassive black hole. M87 can be seen with amateur equipment, but the jet at its center cannot, unless you are a *very* advanced amateur. The galaxy appears to have swallowed some smaller ones. That may be why it's so big. It may have started small and worked its way up. Messier 49 and Messier 84 are two more Virgo Cluster giant ellipticals that you can observe, and Messier 100 is a large spiral galaxy in the cluster.

Clusters of galaxies exist as far as our telescopes can see. At the limit of current technology in the late 20th century, there are about 150 billion galaxies in the observable universe, but nobody has counted them.

Superclusters, great walls, and cosmic voids

You may think that a large cluster of galaxies, up to 3 million light-years across, would be the end. But deep sky surveys indicate that most or all galaxy clusters are themselves grouped into larger forms, called *superclusters*. The superclusters aren't held together by gravity, but they haven't fallen apart either. They appear to have long filamentary shapes and flat, pancake-like shapes. A supercluster can contain a dozen clusters of galaxies, or hundreds of these clusters, and can be 100 or 200 million light-years long.

We are in the outer parts of the Local Supercluster, sometimes called the Virgo Supercluster, which is centered near the Virgo Cluster of Galaxies.

The superclusters seem to be positioned on the edges of huge empty regions of the universe, called cosmic voids. The nearest one, the Bootes Void, is about three million light-years across. Lots of galaxies sit on its periphery, but only a few, mostly little ones, have been found inside it.

Astronomer Robert Kirschner discovered the Bootes Void. But when he was congratulated on the find, he reportedly said modestly, "It's nothing."

Some of the biggest superclusters, or groups of superclusters, are called Great Walls. The first one that was found is about 750 million light-years long. But other Great Walls, far out in the universe, may even be larger. As far as astronomers know, the Great Walls don't display any Great Graffiti, but they have a lot to tell us about the origin of large structures in space and the early history of the universe, if we only understood the language.

Galactic Images on the World Wide Web

This section completes our quick tour of some of the finest sights in the Milky Way and beyond (including one in the Large Magellanic Cloud).

You can find panoramic maps of the Milky Way's galactic plane — as recorded with radio telescopes, X-ray, and gamma-ray observatory satellites, and in visible light (click on Optical) — at the NASA Web site at adc.gsfc.nasa.gov/mw/milkyway.html.

For some of the best color images of nebulae ever photographed, see those of the Space Telescope Science Institute at three different Web pages:

✔ The collection of nebular images originally distributed with press releases is at `oposite.stsci.edu/pubinfo/nebulae.html`.

✔ The Gallery of Planetary Nebula Images is at `oposite.stsci.edu/pubinfo/pr/97/pn`.

✔ The Hubble Heritage project's The Gallery Pages (with wonderful images of galaxies and other objects, too) is at `heritage.stsci.edu/public/gallery/galindex.html`.

Our home planet, Earth and its Moon.

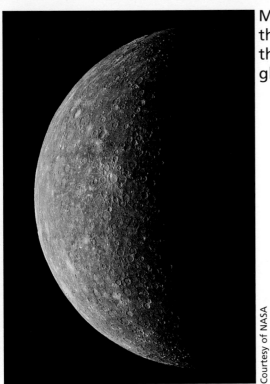

Mercury, the planet closest to the Sun, is frequently invisible to the naked eye, lost in the Sun's glare.

Courtesy of NASA

Courtesy of NASA

Venus, covered in clouds, is the second brightest object in the night sky, after the Moon.

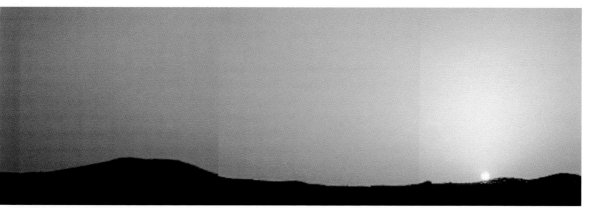

A Martian sunset.

View from the Mars Pathfinder landing site.

Mars is likely to be the first planet visited by people from Earth.

Courtesy of NASA

Jupiter and its four Galilean moons: Io, Europa, Ganymede, and Callisto. This planet has 12 known smaller moons and a ring system.

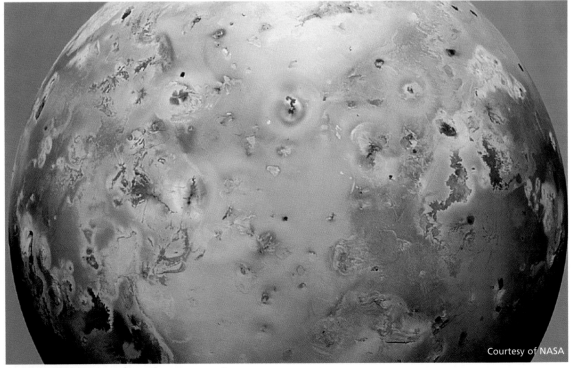

Courtesy of NASA

Jupiter's very volcanic moon, Io.

Courtesy of NASA

Courtesy of NASA

Saturn and two of its 18 known moons.

Courtesy of NASA

False color is used in many space photos to clarify details, indicate composition, or distinguish the regions of such cosmic phenomena as Saturn's rings.

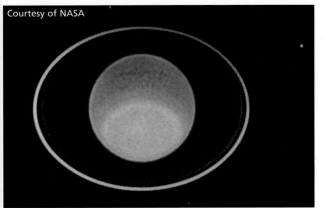

Courtesy of NASA

Like Saturn, Uranus has rings, but they can't be seen with your home telescope.

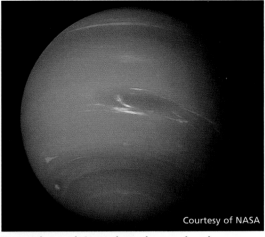

Courtesy of NASA

Streaky white clouds and a large dark spot marked Neptune's atmosphere when this picture was taken.

Courtesy of NASA

One of Neptune's eight known moons, Triton is larger than the planet Pluto and has intriguing surface features.

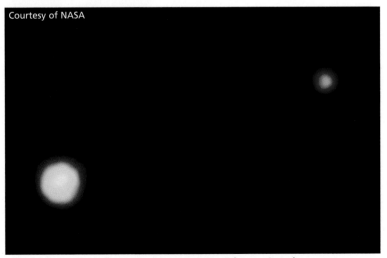

Courtesy of NASA

Clinging to planetary status, Pluto is about two-thirds the size of Earth's moon. Pluto's moon Charon is half the size of Pluto.

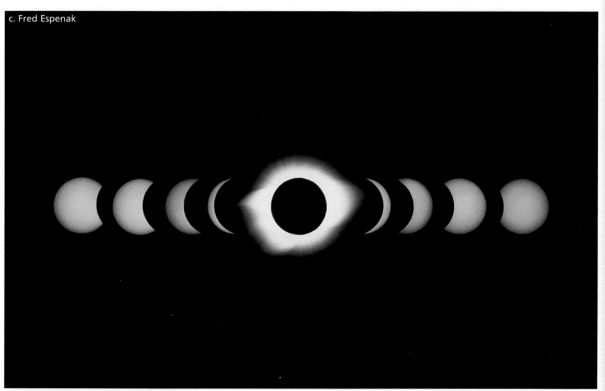

c. Fred Espenak

Eclipsing in stages, 1994

c. Fred Espenak

A total eclipse of the sun, 1998. The last total solar eclipse of the century took place August 11, 1999. The next one won't occur until 2001.

The Large Magellanic Cloud: an irregular galaxy near the Milky Way visible to the naked eye from the Southern hemisphere.

c Anglo-Australian Observatory/Royal Observatory, Edinburgh. Photograph by David Malin

c. 1999 Jerry Lodriguss

The spiral galaxy in Andromeda with its two much smaller companion galaxies.

c. Malin/IAC/RGO. Photograph by David Malin

This spiral galaxy is seen from its side, or "edge-on."

Courtesy of Peter Challis

A bright Type1A supernova (lower left) in a faraway galaxy.

Courtesy of NASA

A young star cluster alongside an emission nebula. Hot young stars energize the nebula and make it glow.

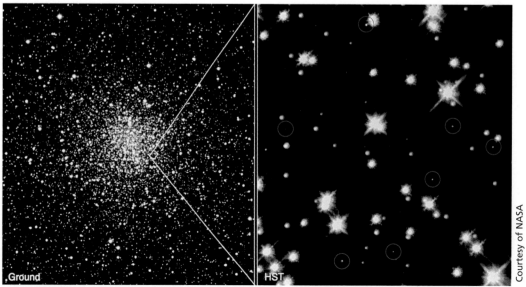

Ground

HST

Courtesy of NASA

Thousands of stars make up a globular cluster; tiny white dwarfs are circled at right.

c. 1999 Jerry Lodriguss

The Pleiades, or Seven Sisters, in Taurus is the best known open star cluster. Star clusters are groups of stars bound together by gravity. Open clusters can have dozens of stars forming no particular shape.

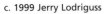
c. 1999 Jerry Lodriguss

The Double Cluster in Perseus is a pair of open clusters of very young stars.

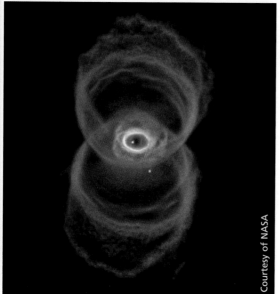

The Etched Hour Glass Nebula is a planetary nebula with a dumbbell structure.

Courtesy of NASA

The Stingray Nebula may be the youngest known planetary nebula.

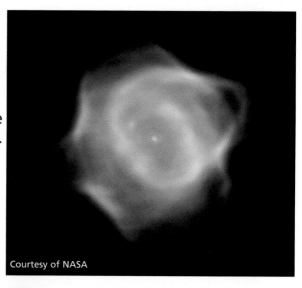

Courtesy of NASA

c. 1999 Jerry Lodriguss

The Trifid Nebula in Sagittarius.

A star cluster (lower right) in the Tarantula Nebula, a star nursery deep in the

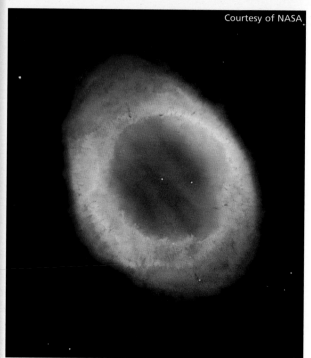

Courtesy of NASA

The Ring Nebula is the last gasp of a
dying sun. Nebulas are associated with
both the birth and the death of stars.

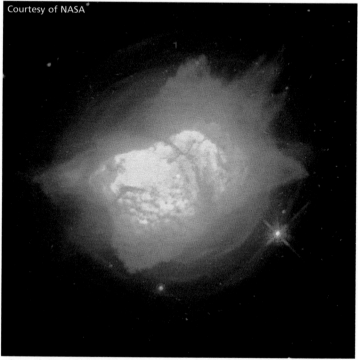

Courtesy of NASA

The dense planetary nebula NGC 7027 hides
its central star from view.

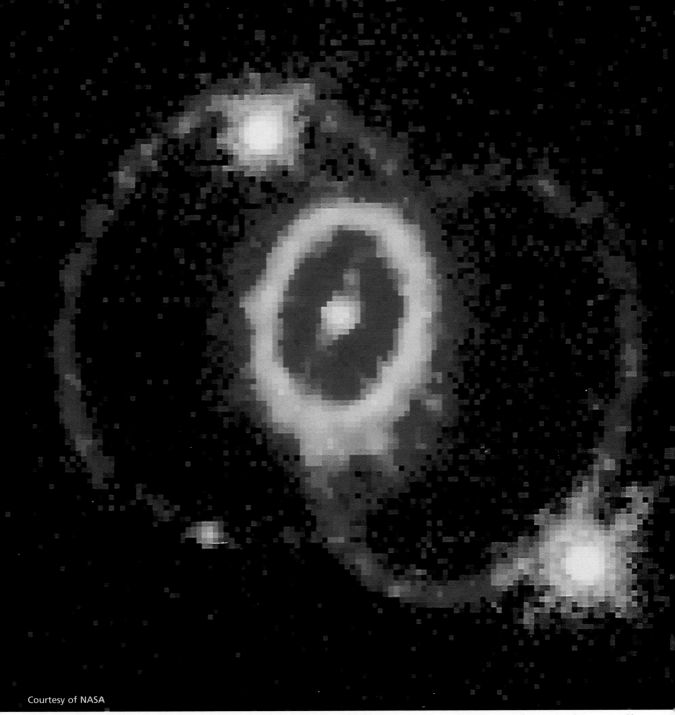

The expanding remains of the exploded star Supernova 1987A make up the small central blob in this image. Around the blob are a bright central ring and two outer rings, all composed of gas that left the star long before the star exploded.

A huge planet circles closely around the star 51 Pegasi in this artist's concept of an extra solar planet that has been detected but not photographed.

Chapter 13

Black Holes and Quasars

- -

In This Chapter
▶ Digging deeper into black holes
▶ Knowing your quasars from your blazars
▶ Unifying active galactic nuclei

- -

Black holes and quasars are two of the most exciting and sometimes mystifying areas of modern astronomy, and it turns out that the two subjects are related. In this chapter, I explain the connection between them.

You may never see a black hole through your own telescope. But I can guarantee that when people learn you are an astronomer, they will ask you "What's a black hole?" I mention black holes briefly in Chapter 11, but in this chapter I offer you the "hole" treatment.

Black Holes: Weird and Irresistible

You can fall in a black hole, but you can't fall out — you can't even get out if you want to (and you *would* want to). You can't even call home. ET was lucky he landed on Earth, not in a black hole, because he was still able to make the call.

A *black hole* is an object in space whose gravity is so powerful that not even light itself can escape from within — which is why black holes are invisible.

Anything that enters a black hole needs more oomph than than it can ever have in order to get back out. The formal name for oomph is escape velocity. Rocket scientists use the term *escape velocity* to mean the speed at which a rocket or any other object must travel in order to escape Earth's gravity and pass into interplanetary space. The term applies in a similar way to any object in the universe.

The escape velocity on Earth is 7 miles per second (11 kilometers per second). Objects with weaker gravity have slower escape velocities (the escape velocity on Mars is only 3 miles per second, or 5 kilometers per second), and objects with more powerful gravity have higher escape velocities. On Jupiter, the escape velocity is 38 miles per second (61 kilometers per second). But the *world* titleholder for escape velocity will always be a black hole. The gravity of a black hole is so strong that its escape velocity is greater than the speed of light (186,000 miles per second, or 300,000 kilometers per second). Nothing, not even light, can escape from a black hole. (Since you need to travel faster than the speed of light to escape a black hole, and nothing — including light — travels faster, there's no way for light to get out.)

Types of black holes

Scientists are able to detect black holes when we see gas swirling around them that's too hot for normal conditions. We detect jets of high energy particles making their escape as though to avoid falling into the black hole, and we even detect stars racing around orbits at fantastic speeds, as though driven by the gravitational pull of an enormous unseen mass (which they are).

As I mention in Chapter 11, two main types of black holes are recognized: stellar mass black holes, which have the mass of a normal star, and supermassive black holes, which are almost a million to a few *billion* times more massive than the Sun.

Intermediate black holes, which have masses 500 to 1,000 times that of the Sun, were discovered in 1999. Their role in the universe is even less well understood than that of the stellar mass and supermassive varieties.

What's inside black holes?

A black hole has three parts:

- The event horizon, which is the perimeter of the black hole
- The singularity, the heart of the hole formed from the ultimate compression of all matter within it, except:
- Stuff that is falling from the event horizon toward the singularity

The following sections describe these parts in more detail.

The event horizon

The event horizon is a spherical surface that defines the black hole (see Figure 13-1). Once inside the event horizon, an object can never get back out of the black hole or be visible again to anyone on the outside. And from inside that horizon, nothing on the outside can be seen either.

A Black Hole

Figure 13-1:
One concept of a black hole. Arrows represent doomed matter falling in.

The size of the event horizon is proportional to the mass of the black hole. Make the black hole twice as massive, and its event horizon will be twice as wide. If scientists had a way to squeeze Earth down into a black hole (we don't, and if we did, I wouldn't tell how), it would have an event horizon less than ¾ inch (about 2 centimeters) across.

Small or *stellar mass black holes* have masses from about three times the mass of the Sun on up. *Supermassive black holes* have hundreds of thousands or even a few billion times the mass of the Sun. Stellar mass black holes result from the deaths of large stars, as I describe in Chapter 11. Supermassive black holes appear to be at the centers of galaxies and may come from the merger of many closely packed stars around the time the galaxies formed. But nobody knows for sure.

Table 13-1 offers a list of black hole sizes, in case you'd like to try some on.

Table 13-1	Black Hole Measurements		
Black Hole Mass (in Solar Masses)	**Diameter of Black Hole in Miles**	**Diameter of Black Hole in Kilometers**	**Comment**
3	11	18	Smallest stellar mass black hole
10	37	60	
100	370	600	Largest stellar mass black hole
1000	3700	6000	Intermediate mass black hole
1 million	3.7 million	6 million	Black hole at Milky Way center
1 billion	3.7 billion	6 billion	Black hole in a quasar

As far as we know, no existing black holes are smaller than about 3 solar masses and eleven miles wide.

The singularity and falling objects

Anything that falls inside the event horizon moves down toward the singularity. There it merges into the singularity, which scientists believe is infinitely dense. We don't know what laws of physics apply at the immense densities that are reached near or in the singularity, so we can't describe what conditions there are like. This is literally a "black hole" in our knowledge.

Some mathematicians think that at the singularity, there may be a *wormhole,* a passage from the black hole to another universe. The wormhole concept has lured authors and movie directors to produce lots of science fiction on the topic. But they're just fishing. (One of the latest books is *Cosm* — by Gregory Benford and published by Avon Books, 1998 — in which physicists on Long Island make a wormhole.) Most experts think that wormholes don't exist. And even if they do, we have no way to see them inside black holes, nor to worm our way down to them. Another theory says that at the location where a hypothetical wormhole connects to another universe, there is a *white hole,* a place where enormous energy pours out into the other universe, a gift from ours. This idea, too, seems to be wrong, but even if the theory is correct, we would have to journey to another universe (talk about frequent-flyer miles!) to see one.

Traveling to another universe is out of the question (at least for now). But the other possibility, of course, is to look for white holes in our universe, where wormholes from other universes might emerge. But scientists can find no such thing. Someone once suggested that quasars may be wormholes. But now astronomers have perfectly good explanations for quasars (as I explain in the section, "Quasars: Defying Definitions," later in this chapter), so as far as I am concerned, astronomers are off the hook.

What's on the outside of black holes

In actual cases of celestial objects that scientists believe are harboring black holes, the following is sometimes what seems to be happening:

1. Gaseous matter falling toward the black hole swirls around in a flattened cloud called an *accretion disk*.

2. The closer the gas in the accretion disk gets to the black hole, the denser and hotter the gas gets.

   The gas heats up because it's being squeezed by the gravity of the black hole, a process that occurs because friction increases as the gas gets denser. (The process resembles the way air conditioners and refrigerators work: When gas expands, it gets cooler; but when gas is compressed, it gets hotter.)

3. As the gas approaches the black hole and heats up, it glows brightly. This radiation from the accretion disk can take many forms, but the most common form is X rays. X-ray telescopes such as NASA's newest large orbiting observatory, CHANDRA, detect these X rays and allow scientists to pinpoint the black hole.

   So although you can't actually see a black hole through a telescope, you can detect radiation from the accretion disk that is swirling around it — if you have an X-ray telescope and you are floating in space. X rays don't penetrate Earth's atmosphere so astronomers use telescopes located in space to observe them.

Bare black holes may exist out in space, with no gas swirling into them. If that's the case, astronomers can't see them, unless they just happen to pass right in front of a background star or galaxy while we are looking. In that case, we might infer that the black hole exists because we see the effect of its gravity on the appearance of the background object. But this situation would be a rare coincidence. Don't hold your breath waiting to spot a bare black hole.

Distortions of space and time

Another way of describing a black hole is as a place where the fabric of space and time are highly distorted. A *straight line* — which is defined in physics as the path taken by light moving through a vacuum — becomes curved in the vicinity of a black hole. And, as an object approaches a black hole, time itself behaves oddly, at least as perceived by an observer at a safe distance.

Suppose that you are at a safe distance, but have launched a robotic probe into a black hole. A big electric billboard on the side of the probe displays the time as given by an onboard clock.

You watch the clock through a telescope in your spaceship as the robotic probe falls toward the black hole. You see that the clock runs slower and slower as the probe approaches the black hole. In fact, you can never actually see the probe fall in. You will see it get redder and redder as its glow is red-shifted by the powerful gravity of the black hole. After a while, the glow of the electric billboard will be redshifted to infrared light, which your eyes can't detect. (See Chapter 11 for a discussion of the Doppler Effect and redshifting.)

Now consider what you would see if you were on board the falling probe. (Don't try this at home; in fact, don't try it anyplace.) You can watch the face of the clock inside the probe, and peek back where you've come from through a window. You, the ill-fated onboard observer, see that the clock runs perfectly normally. You don't perceive that it runs slow at all. As you look out the window at the mother ship and the stars, everything seems to be blueshifted. You're blue at the thought that you can never go home again. You pass through an invisible boundary around the black hole in no time at all. The boundary is the event horizon, and once inside this horizon, you can never see the outside again, nor can anyone outside see you.

As seen from the mother ship, you never enter the black hole; you just get closer and closer. But on the falling probe, you can tell that you've dropped right in. At least, that's the case if you are still alive. Ultimately, anyone who falls into a black hole is pulled apart by tidal force, an effect of the black hole's immense gravity. At least, along one dimension, you are pulled apart. To make things worse, in the other two spatial dimensions, tidal force squeezes you together unmercifully.

If you enter the black hole feet first, you are stretched out (if you're not already pulled apart) until you are tall enough to be drafted as a center in the National Basketball Association. But from bellybutton to back and from hip to hip, you will be squeezed together like coal turning into diamond under immense pressure inside Earth. Only worse.

Small or stellar mass black holes are the most deadly ones, just as some small spiders are more poisonous than big tarantulas. If you fall toward a stellar mass black hole, you are torn apart and squeezed together before you enter, and you never get to see the universe disappear before you are done for. But falling into a supermassive black hole is a happier experience. You get to fall inside the event horizon and see the universe black out before you suffer the tidal fate (or is it the fatal tide?).

Considering that black holes are all around us in the universe, you can see why scientists want to spot them and study them, but keep a safe distance.

Quasars: Defying Definitions

There are at least two definitions of quasars, the original definition and the current definition:

✔ **The original definition:** *Quasar* is an abbreviation or acronym for "quasi-stellar radio source." That term means a celestial object that emits strong radio waves, but looks like a star through an ordinary visible light telescope (see Figure 13-2).

There's nothing wrong with the original definition of quasar except that it turns out that, at most, 10 percent of all objects that we now call quasars fit this definition. The other 90 percent don't make strong radio waves. They are what astronomers call radio-quiet quasars.

✔ **The current definition:** A *quasar* is a bright object at the center of a galaxy that produces about 10 trillion times as much energy per second as the Earth's Sun, and whose emissions are highly variable at all wavelengths.

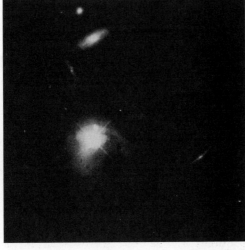

Figure 13-2:
A quasar in a seemingly disrupted galaxy shines at lower left of center.

Courtesy of NASA

After decades of puzzling over what quasars are, astronomers have concluded that they represent giant black holes at the centers of galaxies. Enormous energy is released by matter that falls into the black holes, and the observed energy sources are what astronomers call quasars.

All quasars produce strong X rays; about 10 percent produce strong radio waves, and they all emit ultraviolet, visible, and infrared light, too. All of those emissions can vary over weeks, months, years, and even over times as short as one day.

The fact that quasars often change significantly in brightness over the course of a single day indicates to scientists something of surpassing importance: The quasar must be no larger than about one *light-day,* the distance that light travels through a vacuum over the length of a day. And a light-day is only 16 billion miles (26 billion kilometers) long. This number means that a quasar, producing as much light as 10 trillion of Earth's Suns, or 100 times as much light as the Milky Way galaxy, is not very much bigger than our solar system, which is a tiny part of our galaxy.

If the quasar were much bigger than a light-day, it could not fluctuate markedly in such a short time, any more than an elephant could flap its ears as rapidly as a hummingbird flutters its wings.

Quasars that are strong radio sources often display *jets,* long narrow beams in which energy is shooting out of the quasars in the form of high-speed electrons and perhaps other rapidly moving matter. Often the jets are not smooth, but lumpy, with blobs of matter moving outward along the beams. Sometimes the blobs appear to be moving at more than the speed of light. This *superluminal motion* is an illusion related to the fact that the jets in such cases are pointing almost exactly at Earth; the matter in them is actually moving close to light speed, but not faster than light.

Many books say that a quasar has very broad lines in its spectrum, corresponding to red and blue shifts of gas roiling within the quasar at up to 6,000 miles per second (10,000 kilometers per second). This statement is not always true. Quasars come in a variety of types, and some don't have these broad spectral lines.

But the broad spectral lines are an important trait of many quasars and a clue to their relationship to other objects, as I describe in the next section.

Active Galactic Nuclei: What in Blazes Are Blazars?

For years after quasars were discovered, astronomers argued about whether they were located in galaxies. Today, we know that they are, but that's because technology has improved to the point that we can make a telescopic image that shows both a quasar and the galaxy around it. The latter is called the quasar's *host galaxy*. Because a quasar can be 100 times brighter than its host galaxy, or even brighter, hosts tend to be lost in the glare of their quasar guests, like the homeowner who puts up a presidential candidate for a night.

Electronic cameras, which can record a greater range of brightnesses in a single exposure than photographic film, made the discovery possible.

Quasars are an extreme form of what astronomers now call *active galactic nuclei (AGN)*. The term designates the central object of a galaxy when the object has, well, quasarlike properties, such as a very bright starlike appearance, very broad spectral lines, and detectable brightness changes.

These are the main terms used to describe active galactic nuclei (AGN):

- **Radio-loud quasars ("original quasars") and radio-quiet quasars (90 percent or more of quasars):** These are the quasars the preceding section describes. They are similar kinds of objects, with and without strong radio emission. They are located in spiral galaxies, like the Milky Way. No quasar is visible in the Milky Way galaxy, but there is evidence of a roughly 1-million-solar-mass black hole at the galaxy's center.

- **Quasistellar objects (QSOs):** This term applies collectively to the radio-loud and radio-quiet quasars. Some astronomers just lump them all together as *QSOs*.

- **Seyfert galaxies:** These spiral galaxies have an AGN at their centers. A Seyfert AGN is a lot like a quasar, with broad spectral lines, and rapid brightness changes. It may be as bright as the host galaxy, but not 100 times brighter like a quasar. So the host is not lost in the Seyfert nucleus's glare.

  A Seyfert nucleus is not a demanding guest; it's like a minor presidential candidate who visits a small town in Iowa without causing a big fuss. Local folks know that the candidate is around, but they are not disturbed from their daily routines. Carl Seyfert was an American astronomer who pioneered in studying these galaxies and their bright centers.

- **OVVs:** This term stands for *optically violently variable quasars*. They are quasars with jets that point right at Earth, and which are observed to undergo even more pronounced rapid brightness changes than ordinary, garden-variety quasars. Think of some firefighters struggling to direct a firehose at someone whose clothes are on fire. The water pressure may be unstable, with the water pulsing a bit. As visible by spectators from the side, the stream from the hose may look pretty steady, but the person on the receiving end feels every fluctuation in the flow as the oncoming water batters him. OVVs are the firehoses of quasardom.

✔ **BL Lacs:** This term is astronomy lingo for *BL Lacertae objects*. BL Lacs as a group are AGN that resemble BL Lacertae. BL Lacertae changes in brightness and for years was thought to be just another variable star in the constellation Lacerta (it looks like a star in photographs of the sky). Then it was identified as a strong source of radio waves and eventually it was pinned down as the active nucleus of a host galaxy that had been lost in its glare.

Unlike most quasars, a BL Lac doesn't have broad spectral lines. And its radio waves are more highly polarized than those from ordinary radio-loud quasars. *Polarized* means that the waves have a tendency to vibrate in a preferred direction as they travel through space. Unpolarized waves vibrate equally in all directions as they move. At the ballpark, you can't tell the players without a scorecard and at the observatory, you have to check polarization to know your radio-loud quasars from your BL Lacs.

✔ **Blazars:** These are OVVs and BL Lacs. It's just a term meant to cover both types of object. OVVs and BL Lacs have many similarities. Both are highly variable in brightness and their jets are thought to point right at Earth. And they are all radio-loud.

Do we really need the term "blazars?" I'm not so sure. My friend Dr. Hong-Yee Chiu became famous among scientists for coining "quasar." His friend, Professor Edward Spiegel, invented "blazar" a few years later. If you discover a new kind of object, or write one of the leading studies of it, you may get to name it, too. Adding "ar" to your own name is not allowed; the term should be descriptive of the scientific properties of the object, not the astronomer.

✔ **Radio galaxies:** These are galaxies with active galactic nuclei that are not especially bright but produce strong radio emission. Most of the strongest radio-emitting galaxies are giant elliptical galaxies. Often, they have beams or jets that transport energy from the AGN to huge lobes of radio emission, empty of stars, far outside and far larger than the host galaxy itself. There are usually two lobes, on opposite sides of the galaxy.

All these different types of active galactic nuclei have one thing in common: They are powered by energy somehow generated in the vicinity of a super-massive black hole at their center.

Near the supermassive black hole, stars orbit the center of the host galaxy at immense speeds. That's how astronomers measure black holes' masses. With telescopes such as the Hubble, they determine the velocities of the orbiting stars, or sometimes, orbiting gas clouds. They do this by measuring Doppler shifts of the emission from the stars or the gas. The speeds indicate the mass of the central object. If the black hole were less massive, the stars at a given distance from the center would orbit at a slower pace.

In a quasar or a radio galaxy of the giant elliptical type, the black hole often attains a billion solar masses, or even a few times more. In Seyfert galaxies, the black hole mass is often around a million solar masses.

The black hole makes it possible for the AGN to shine, but only the mass falling into the black hole actually powers the shine. To make a quasar shine may take 10 times the mass of our Sun per year, swirling in toward the black hole.

If there's no material falling into the black hole, it will not reveal itself by producing a bright glow, radio emission, or strong X rays. Like kids who depend on their school lunches, the black holes only shine when they are fed. Supermassive black holes may be lurking at the centers of most galaxies, but in most cases, they are not being fed. So astronomers only see quasars or other kinds of AGN in a small fraction of galaxies.

The *Unified Model of Active Galactic Nuclei* is a theory that proposes that all AGN are the same, but that astronomers are looking at them from different directions with respect to their accretion disks and their jets. Also, the black holes are being fed at different rates, so some AGN are brighter than others for that reason alone. Dozens of astronomers write papers on the Unified Model every year, some finding evidence for and some finding evidence against.

I think that there are real differences among the different types of AGN, but that they have many basic similarities. Astronomers need more information before we can unite around the Unified Model or any other theory of AGN. In the meantime, what do you think? Your taxes pay for much of this research, and you are entitled to an opinion.

Part IV
The Remarkable Universe

The 5th Wave   By Rich Tennant

"Along with 'Antimatter' and 'Dark Matter,' we've recently discovered the existence of 'Doesn't Matter,' which seems to have no effect on the universe whatsoever."

In this part . . .

Read this part when you need a diversion, something to stir your mind with thought-provoking ideas and possibilities. Curl up with a cup of cider and read about SETI, the search for extraterrestrial intelligence. Have scientists found any evidence that those little green men are out there? Find out about dark matter and antimatter (yes, antimatter exists in the real world, not just in science fiction). And, when you are ready, ponder the entire universe: How did it begin, what shape is it, and what's going to happen to it?

Chapter 14

SETI and Planets of Other Suns

● ●

In This Chapter

▶ Making a case for extraterrestrial intelligence

▶ Participating in SETI projects

▶ Hunting for extra-solar planets

● ●

The universe is both vast and varied. But do we share these starry realms with other thinking beings? Anyone who tunes in *Star Trek* or frequents the local cineplex already knows Hollywood's answer: The cosmos is cluttered with aliens (many of whom have managed to pick up a goodly amount of uninflected English).

But what do scientists say? Are there really extraterrestrial beings out there? Most researchers believe that the answer is yes. Some of them are even looking for evidence. Their quest is known as SETI (rhymes with "yeti"), the Search for Extraterrestrial Intelligence. (Other scientists are searching or planning to search for traces of primitive life on Mars, but SETI seeks advanced civilizations that are capable of broadcasting into space.)

Is Anybody Out There?

Why are many scientists optimistic about the possible existence of aliens?

Most of their upbeat attitude derives from the fact that our place in the cosmos is thoroughly unremarkable. The Sun may be an important star to us, but it's a bit player in the universe. The Milky Way Galaxy hosts ten billion similar suns. And if this number fails to impress you, note that over a hundred billion other galaxies are within range of our telescopes. The bottom line is that far more sunlike stars are sprinkled through the visible universe than there are blades of grass on Earth. To assume that our grass blade is the only one where something interesting happens is (to put it gently) a bit gutsy. Distressing as it may be to our self-esteem, this planet may not be the intellectual nexus of the universe.

How might Earthlings find these brainy brethren? We can't go visit their likely homes. Rocketing off to distant star systems, although a work-a-day staple of science fiction, is actually quite difficult. The speed of earthly rockets, an impressive 30,000 miles per hour, is less impressive when you reckon that it would take these craft a thousand centuries to reach Alpha Centauri, the nearest stellar stop on our tour of the universe. Faster rockets would take less time, but would consume more energy — a *lot* more energy.

SETI and Drake's Equation

Although Earthlings can't go visit, we might be able to find evidence of technically sophisticated space aliens by eavesdropping on their radio traffic.

In 1960, astronomer Frank Drake attempted to listen in on cosmic communications by using an 85-foot-in-diameter radio telescope in West Virginia. If you've seen the movie *Contact,* you know that a radio telescope is similar to a backyard satellite dish that's been seriously bulked up (see Figure 14-1). Drake connected his antenna to a new, sensitive receiver working at 1,420 MHz (located in what's called the microwave region of the radio spectrum), and then pointed the telescope at a couple of Sunlike stars.

Figure 14-1:
A radio telescope.

Photo by Seth Shostak

Drake didn't hear any aliens with his Project Ozma, but he provoked a great deal of enthusiasm within the scientific community. A year later, in 1961, the first major conference on SETI was held, and Drake tried to organize the meeting by distilling all the unknowns of the search into a single equation, now known as the *Drake Equation*. (For the mathematically inclined, I provide this simple little formula in the sidebar, "The Drake Equation.") Its logic is easy. The idea is to estimate *N*, the number of civilizations in our galaxy that are using the radio airwaves now. This number clearly depends on the number of suitable stars in the galaxy, times the fraction that have planets, times the number of . . . Well, you can read about it in the sidebar.

TECHNICAL STUFF

The Drake Equation

Frank Drake's nifty little formula is often used as the basis for discussions about SETI and the chances that human beings will ever make contact with extraterrestrial intelligent life. The equation is quite simple and doesn't require any math beyond what you were supposed to have mastered in the eighth grade.

The equation computes *N*, the number of broadcasting civilizations active in the Milky Way galaxy. As with the Bible, several versions of Drake's Equation are in use, but here's the usual formulation, in all its gory glory:

$$N = R^* f_p\, n_e\, f_l f_i f_c\, L$$

R^* is the rate at which long-lived stars, suitable for hosting habitable planets, are formed in the galaxy. Because the Milky Way has roughly 400 billion stars and is approximately 10 billion years old, this number is about four per year.

Remember: About one in ten stars is close enough in size and brightness to our Sun to be considered suitable to have orbiting habitable planets.

f_p is the fraction of good stars that have planets. No one knows what this number is, but it's at least 3 percent and could be much higher.

n_e is the number of planets per solar system that are capable of incubating life. In our own solar system, that number is at least one (Earth), but in someone else's system, no one knows. A typical guess is one.

f_l is the fraction of habitable planets that actually develop life. It's not unreasonable to assume that most of them do.

f_i is the fraction of planets with life that develop intelligent life. This number is controversial, of course, because intelligence may be a rare accident in biological evolution.

f_c gives the fraction of intelligent societies that invent technology, and in particular radio transmitters. Probably most of them do.

L, the final term, is the lifetime of societies using radio. This term is a matter of sociology, not astronomy, of course, so your guess is as good as the author's. Maybe better.

The value of *N* computed from Drake's Equation depends on your choice of values for the various terms. Pessimists think that *N* may be only one (we're alone in the Milky Way galaxy). Carl Sagan figured it was closer to a million. And what does Drake himself say? "About ten thousand." Moderation in all things.

Drake's Equation is truly seductive, and you may want to impress strangers by rattling it off at a dinner party. But although scientists may know or can safely guess at the values of the first few terms in the equation (such as the rate at which stars capable of hosting planets are formed and the fraction of such stars that actually have planets), we don't have any real knowledge of such things as the fraction of life-bearing planets that develop intelligent life, or the lifetime of technological societies. So Drake's Equation still doesn't have any "answer." It's just a great way to organize the SETI discussion.

Current SETI Searches: Listening for E.T.

Nearly all the modern SETI (Search for Extraterrestrial Intelligence) efforts follow in Frank Drake's footsteps. In other words, they use large radio telescopes in an attempt to eavesdrop on signals from alien civilizations.

Unlike light waves, radio waves easily punch through the clouds of gas and dust that fill the space between the stars. In addition, radio receivers can be quite sensitive. The amount of energy required to send a detectable signal from star to star is no more than is pumped out by your local TV station, assuming that the aliens wield a transmitting antenna a few hundred feet in size.

Project Phoenix

The most sensitive of the modern-day SETI searches is Project Phoenix, run by the SETI Institute in Mountain View, California. This project is a successor to a NASA SETI program that was stopped by Congress in 1993 (indeed, ever since that date, all SETI experiments in the United States have been privately funded).

Project Phoenix is the only large-scale SETI experiment that scrutinizes individual stars. Other projects use their telescopes to sweep large tracts of the sky. Of course, those broad sweeps allow scientists to examine more of the heavens, but by concentrating only on nearby, Sunlike stars, Phoenix can achieve much greater sensitivity. In other words, it could find far weaker radio signals. This search is currently being made with the 1,000-foot-diameter Arecibo radio telescope (in Puerto Rico), the mother of all radio telescopes (see Figure 14-2).

Figure 14-2:
A view of
the massive
Arecibo
radio
telescope.

Photo by Seth Shostak

Phoenix (and many other SETI experiments) looks for signals in the microwave region of the radio dial. Microwaves, aside from their ability to make leftovers palatable, are the preferred "hailing channel" for the SETI crowd for two reasons:

- The universe is rather quiet at microwave frequencies — there is less natural static, a fact that E.T. will also appreciate.

- A natural signal generated by hydrogen gas occurs at 1,420 MHz. Because hydrogen is far and away the most abundant element in the cosmos, every alien radio astronomer will know of this natural marker — and might be tempted to get our attention (or the attention of any other civilization in space) by sending out a signal near to its position on the dial.

But let's face facts. Scientists really don't know *exactly* where the extraterrestrials might tune their transmitters. So Project Phoenix checks out many millions of channels at once (billions of channels for each star that is searched). Figure 14-3 shows part of a SETI receiver that's used by the project.

Figure 14-3:
This SETI
receiver is
primed for
messages
from
non-Earth
beings.

Photo by Seth Shostak

Assuming that researchers do get an interstellar ping, how would we recognize it? The approach of SETI researchers is to look for narrow-band signals (see Figure 14-4).

These are signals that occur at one narrow spot on the radio dial. Narrow-band emissions are of a type that only a transmitter can make. Quasars, pulsars, and even cold hydrogen gas all make radio waves. But their natural static is spread out in frequency — splattered all over the radio spectrum. Narrow-band signals are the mark of transmitters. And transmitters are the mark of intelligence. It takes a brain to build one.

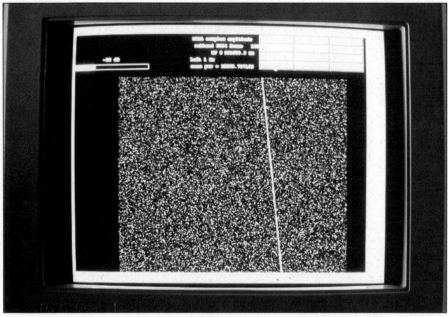

Photo by Seth Shostak

Figure 14-4:
This detection system display screen will enable Phoenix Project staff to pick up signs of intelligence from outer space.

Other SETI projects

Besides Project Phoenix, several other SETI programs are ongoing:

- ✔ The Planetary Society sponsors Projects *BETA (Billion-Channel Extraterrestial Assay)* and *META (Mega-Channel Extraterrestial Assay),* conducted on radio telescopes near Boston and in Argentina.

- ✔ The *Search for Extraterrestrial Radio Emissions from Nearby Developed Intelligent Populations* (nicknamed *SERENDIP*), conducted from the University of California, Berkeley, uses the Arecibo telescope in a "piggy-back" mode. Scientists commandeer a second, unused, receiver on the telescope and simply accept the random bits of sky at which it's pointed. This apparently aimless approach pays off handsomely in observing time: SERENDIP is collecting data almost every day, all day.

- ✔ Another piggyback experiment is *Southern SERENDIP,* run by the SETI Australia Centre, in New South Wales. They use a 210-foot radio tele-scope at Parkes, in the sheep and mosquito country a few hundred miles west of Sydney.

- ✔ Also, the SETI League, headquartered in scenic New Jersey, is recruiting radio amateurs to use their backyard dishes in the search for cogitating aliens.

All the major SETI programs have their own Web sites. You can find links to them at the SETI Institute's site at `www.seti.org` or at the Planetary Society's Web page at `seti.planetary.org`.

SETI searchers want you!

Another Web address you may want to laboriously copy from this book is that of the SETI@home project: `setiathome.ssl.berkeley.edu`.

SETI@home is a part of project SERENDIP. If you go to its site, you can download a really snazzy screensaver, without cost. Once the software is installed on your desktop computer, your modem will connect to a server at Berkeley to retrieve a chunk of SETI data. Then the screensaver software will crunch away on these data, looking for signals. After a few days (depending on how often you leave your computer to its own devices), the results will be uploaded back to the server.

Although your chances of finding E.T.'s telltale tone are small, they're not zero. Who knows? You might enjoy sharing a plate of meatballs with the Swedish king after collecting your Nobel Prize.

Hot Jupiters: The Truth about Extra-Solar Planets

One term of Drake's famous formulation is f_p, the fraction of sunlike stars that sport planets. Astronomers have believed for decades that planets are plentiful, simply because the birth of a star is inevitably accompanied by leftover material — a messy residue of gas and dust that could turn into small, orbiting worlds.

But to actually find planets around stars is tough. You can't simply point a telescope in the direction of a nearby star and hope to see its planets. They're too dim and too close to a blinding light source (their sun). To grasp the full, existential challenge of the problem, imagine trying to see a marble located 30 yards from a light bulb at a distance of 10,000 miles.

Despite these daunting difficulties, astronomers *have* found *extra-solar planets* (planets outside of our solar system, which are orbiting stars other than our Sun), not by picking them out in photos, but by measuring the motions of their host stars.

Planets and stars orbit their common center of mass, and this arrangement means that both objects move. As they swing around under the influence of their mutual gravitational attraction, the star pulls on the planet, making the planet move, and the planet pulls on the star, making the star move. The planet is a lot less massive than the star, so the so-called reflex motion of the star is usually not much — perhaps only 50 miles an hour (compared with the planetary motion, which could be 10,000 miles an hour or more). But using sensitive spectroscopes on large telescopes, astronomers have sought the small Doppler Effect (see Chapter 11) that the slow stellar wobble would produce in the star's light. And they've already managed to find several dozen stars whose lazy dance betrays orbiting planets.

51 Pegasi's warm little world

The first extra-solar planet to be found around a normal star was announced in the fall of 1995 by two Swiss astronomers, Michael Mayor and Didier Queloz. The discovery caused a great deal of consternation in the research community, mainly because the new planet whipped about its star (51 Pegasi) at a breakneck pace. A complete orbit took only about 4 days. Consequently, this planet is known to lie a trifling 5 million miles from its host star (see Figure 14-5). That's about eight times closer than Mercury is to the Sun, and implies that the temperature of this new-found world is roughly 1,000° C. The size of 51 Pegasi's stellar wobble indicates that its planet's mass is at least half that of Jupiter. For obvious reasons, the new planet was soon dubbed a _hot Jupiter._

Figure 14-5:
An artist's concept of how close the new planet must be to its sun.

Photo by Seth Shostak

In the 4 years following the discovery of 51 Pegasi's warm little world, approximately two dozen other planets were found, nearly all by spectroscopic measurement of Doppler shifts. Quite a few of these are also hot Jupiters — massive planets that hug their suns tighter than a loving mom.

But it's unlikely that any of these hot and heavy worlds were born in their present, toasty orbits. It's far easier to form large planets in the dim suburbs of a solar system. The colder temperatures and endless expanse of material in these nether regions encourage the rapid conglomeration of icy debris into large worlds. But once born, these planets' interactions with the leftover debris could cause them to wander from home and gravitate inward to the hellish domains of their own, scorching suns.

No one is quite sure what prevents these heat-seeking heavyweights from careening into their host stars. One possibility is that these planets raise waves of hot gas on the star's outer surface, and the gravitational effects of these tides halt the planet's inward spiral. But that's still a theory, and astronomers candidly admit that both the birth and ultimate fate of hot Jupiters are phenomena that we don't yet understand.

The Upsilon Andromedae system

In 1999, Geoff Marcy, Paul Butler, and collaborators (who had discovered many of the new planets detected since 1995) added to the planet-finding excitement by claiming that not one, but three large planets were in orbit around the star Upsilon Andromedae. They made this discovery by careful analysis of that star's subtle wobbling motions.

Upsilon Andromedae, an F-type star at 44 light-years' distance from Earth, thus became the first normal star other than the Sun known to have a genuine solar *system*. Once again, the planets themselves are hefty, weighing in at greater than 0.7, 2.1, and 4.6 times the mass of Jupiter. They are not all sun-hugging worlds, however. The outer two have orbits comparable in radius to those of Venus and Mars.

Planets suitable for life?

Although it's reassuring for those who search for extraterrestrials to know that E.T. will have plenty of homes to phone, the new planet discoveries are also a bit disconcerting. After all, hot Jupiters (or, for that matter, cold Jupiters) aren't likely places for biology to cook up. If these oversized planets are typical of the galaxy's complement of worlds, then Earth shouldn't expect much cosmic company.

But that scenario is unlikely to be the case. The technique used to find the planets — looking for the Doppler Effect in the light from stars — is best for uncovering giant planets, close-in to their stars — hot Jupiters, in other words. The search so far might be compared to a reconnaissance of the African savannas from a helicopter. You can see the elephants and rhinos, but would miss the mice and mosquitoes. Scientists have found big planets because we *can* find big planets. Small planets are probably plentiful, but until we build some new types of telescopes, discovering them will be difficult.

If you're keen to know the latest and greatest in the search for extra-solar planets, you can find the facts at `cfa-www.harvard.edu/planets`, which also has links to many other related sites.

Continuing the Search

Although radio searches are, and have been, the darlings of the SETI community, scientists are showing a growing interest in looking for intense light beacons from the stars. Powerful lasers, particularly those that operate at infrared wavelengths, can produce incredibly bright, brief flashes of light. Indeed, these flashes can even outshine the Sun for a trillionth of a second or so (at least at the wavelength of the laser light). Perhaps the aliens are attempting to get our attention by aiming souped-up laser pointers our way. The first tentative attempts at *optical SETI,* as it's called, have already been made.

It's been 40 years since Frank Drake made the first efforts to put us in touch with aliens. So far, not a single confirmed extraterrestrial peep has been snagged by our telescopes. But keep in mind that until now the search has been limited. As technology (and, one hopes, funding) continues to improve, the chance for success increases. Someday soon we might find ourselves puzzling over a signal that comes from the cold depths of space. Maybe it will teach us interesting lessons, such as the meaning of life, or at least all the laws of physics. But one thing is for sure: It will tell us that we have company.

This chapter was contributed by Dr. Seth Shostak, Public Programs Specialist for the SETI Institute in Mountain View, California.

Chapter 15

Dark Matter and Antimatter

. . . .

In This Chapter

▶ Understanding the need for dark matter

▶ Shedding light on the nature of dark matter

▶ Searching for the mystery matter

▶ Getting to know antimatter

. . . .

Stars and galaxies set the night sky aglow, but these glittering jewels account for only a tiny portion of the matter in the cosmos. There's more to the universe than meets the eye — much more.

This chapter introduces you to the concept of dark matter, tells you why astronomers are convinced that the stuff must exist, and describes experiments that may shed light on the nature of this mysterious, invisible material. I also discuss another exotic type of matter in the universe: antimatter. Yes, antimatter exists in the real world, not just in the realms of science fiction. And it's every bit as fascinating as the sci-fi books, television shows, and movies suggest.

Dark Matter: The Glue That Holds Galaxies Together

As far back as the 1930s, astronomers found hints that at least 90 percent of the mass in the universe doesn't emit light.

This invisible material, known as *dark matter,* provides the gravitational glue that keeps a rapidly rotating galaxy from flying apart and enables fast-moving galaxies in a cluster to stick together. Dark matter also seems to have played a crucial role in development of the universe as known today — a spidery network of immensely long superclusters of galaxies separated by giant voids (see Chapter 12). Indeed, dark matter may determine the ultimate fate of the cosmos.

The matter behind the missing mass

The first hint that the universe contained dark matter appeared in 1933. While examining the motions of galaxies within a large cluster of galaxies in the constellation Coma Berenices, astronomer Fritz Zwicky of the California Institute of Technology found that some galaxies moved at an unusually high speed. In fact, these galaxies of the Coma Cluster were moving so rapidly that all the visible stars and gas in the cluster couldn't possibly keep the galaxies gravitationally bound to one another, according to the known laws of physics. Yet somehow the cluster remained intact.

Zwicky concluded that some sort of unseen matter must exist within the Coma Cluster and provide the missing gravitational attraction.

As surprising as this conclusion was, dark matter didn't make headlines for several more decades. Many astronomers figured that, once the motions of galaxies were studied in greater detail, the rationale that predicted the invisible material would disappear. Instead, in the 1970s, evidence for dark matter became more compelling. Not only did star clusters seem to contain the stuff, but so did individual galaxies. The following sections describe the main arguments in favor of dark matter.

Outer stars and inner stars keep pace

Vera Rubin and Kent Ford of the Carnegie Institution of Washington, D.C., were studying the motions of stars in hundreds of spiral galaxies when they came to a result that seemed to fly in the face of conventional physics. A spiral galaxy resembles a flattened fried egg, with most of its mass seemingly concentrated in the yolk — astronomers call this the *bulge* (as Chapter 12 explains). Images reveal that the visible mass of a spiral diminishes rapidly with distance from the bulge.

Scientists would naturally expect that stars in a spiral galaxy would orbit this massive center in the same way that planets in our solar system orbit the Sun. Obeying Newton's law of gravity, the outer planets, such as Pluto and Neptune, orbit the Sun more slowly than do the inner planets, such as Mercury, Venus, and Earth. Therefore, stars in the outskirts of a spiral galaxy should orbit more slowly than those nearer the bulge. But that's not the result Rubin and Ford found.

In galaxy after galaxy, their observations revealed that the outlying stars orbited rapidly, just like the inner ones. With so little visible material in the outer regions, how did the outlying stars manage to zip around so fast and still stay bound to the galaxy? They should have been thrown off at those speeds.

The astronomers concluded that *visible matter* — the stars and luminous gas that show up on telescopic photographs — make up only a small portion of the total mass of a spiral galaxy.

Although the visible mass is indeed concentrated at the center, a vast quantity of other material must extend far beyond. Each spiral galaxy must be surrounded by an immense *halo* of dark matter. And to exert enough of a gravitational tug on the stars in the visible outskirts of the galaxies, the dark matter must exceed the visible matter by at least a factor of 100 in mass. Other types of galaxies (elliptical and irregular) also have dark matter halos.

Cold dark matter accounts for lumps in the cosmos

Cosmologists (scientists who study the large-scale structure of the universe and how it was formed) also have had to invoke dark matter to explain a fundamental riddle about the universe: How did it evolve from a nearly uniform soup of elementary particles in the aftermath of the Big Bang (which I describe in Chapter 16) to reach its present lumpy structure of galaxy clusters and superclusters?

Even though approximately 15 billion years have passed since the birth of the universe, that's not enough time for visible matter on its own to have coalesced into the huge cosmic structures seen today.

To solve this cosmological conundrum, scientists hypothesize that the universe contains a special type of dark matter, *cold dark matter,* that moves more slowly and gathers into clumps more quickly than ordinary, visible matter. Responding to the tug of this exotic material, the ordinary matter formed stars and galaxies within the densest concentrations of this dark matter. This theory would explain why every visible galaxy seems to be embedded in its own dark-matter halo.

The universe is largely uniform

Astronomers believe in dark matter for yet another cosmic reason: The universe, on large scales, looks the same in all directions and has an overall smoothness. This consistency in appearance and smoothness indicates that the universe has just the right density of matter, called the *critical density.* The total amount of visible matter that the universe seems to be endowed with is not nearly enough to achieve critical density. Dark matter would take up the slack. And the exact amount of dark matter present could determine whether the universe will expand forever or collapse in on itself.

Matter is more than 90 percent dark

If the previous arguments are correct, at least 90 percent — perhaps even 99 percent — of the matter in the universe is dark. That's a humbling thought.

The universe we see when we peer through a telescope, or when we look up at the night sky teeming with stars and galaxies, is just a tiny fraction of what's out there. To borrow a nautical analogy, if galaxies are like sea foam, dark matter is the vast, unseen ocean in which they float.

Debating the big question: What's the matter in dark matter?

Okay, plenty of good reasons exist to believe in dark matter. But what the heck is this stuff anyway?

Broadly speaking, astronomers divide the possible kinds of dark matter into two classes: baryonic dark matter and oddball dark matter.

Baryonic dark matter: Lumps in space

Some dark matter might consist of the same stuff that the Sun, planets, and people are made of. This kind of dark matter would be part of the family of baryons, a class of elementary particles that includes the protons and neutrons that are found in the nuclei of atoms.

This *baryonic dark matter* could include chunks of any hard-to-see material, including dust, asteroids, brown dwarfs (failed stars), or white dwarfs (the cold, burned-out cores of Sunlike stars). These lumps of material, sometimes referred to as *MACHOs (massive compact halo objects)* may account for the halos surrounding individual galaxies. But there's not nearly enough of them to account for the development of large-scale structure in the cosmos.

Oddball dark matter

Alternatively, dark matter could consist of an abundance of exotic subatomic particles, dreamed up by physicists, that bear little or no resemblance to baryons. These particles include *neutrinos,* which are known to exist, and others with names such as *axions, squarks,* and *photinos,* that have not been found yet.

During the *Big Bang* — the tremendous outpouring of energy that accompanied the birth of the universe — a zoo of weird, dark-matter particles could have been created and a few may have survived. These include the *axion,* a kind of miniature black hole that is 100 billion times lighter than an electron. Even though axions are featherweights, if enough exist, they could contribute significantly to the cosmic mass. Recent experiments suggest that the neutrino (a particle that was thought to have, perhaps, zero mass) does have a mass and could account for a small portion of the dark matter.

Other candidates for oddball dark matter are heavier — about ten times the mass of the proton — but still insubstantial unless they occur in large numbers. These include the yet-to-be-detected partners of subatomic particles such as *quarks* and *photons,* respectively known as *squarks* and *photinos.* Collectively, such exotica are dubbed *weakly interacting massive particles,* or *WIMPs.*

Searching for Dark Matter

Around the world, physicists are designing sensitive detectors to find the elusive, telltale signals of dark matter. Some are analyzing the subatomic debris created by giant atom-smashing devices, which briefly recreate the extreme heat, energy, and densities that were present in the early universe.

The search techniques have to be innovative. After all, scientists are hunting material that by definition can't be seen and, aside from exerting a gravitational force, doesn't interact with other matter.

WIMPS are shy but leave their mark

Consider the effort to find WIMPs. No container can trap these weakly interacting particles, but scientists can look for evidence that they passed through a detector. When a WIMP whizzes past, it slightly heats up one of the detector's atoms, giving it an extra little kick. These encounters are rare. For a typical laboratory detector, this boost might occur only once in many days.

Unfortunately, cosmic rays, energetic particles that stream in from all directions of space, can mimic the action of a WIMP. To minimize bombardment from cosmic rays, the detector is placed in an underground tunnel. Naturally occurring radioactivity from the walls of the tunnel could also heat up the atoms, so the detector is shielded with lead. To reduce the jiggling of atoms that occurs with increasing vigor at higher temperatures, the detector is cooled to near absolute zero.

MACHOs make a brighter image

Because MACHOs are extended, lumpy objects (think of them as the real Miss and Mr. Universe!), looking for MACHOs is easier. The prime method takes advantage of a mind-bending concept from Einstein's Theory of General Relativity. To wit: Mass distorts the fabric of space and the path of a light wave. This concept means that an object that by chance lies along the line of sight between Earth and a distant star will focus the light from that star, briefly making it appear brighter. The more massive the object, in this case a MACHO, the brighter the star will appear during the alignment.

In effect, the MACHO acts as a miniature gravitational lens, or microlens, bending and brightening the light from the background star. (See Chapter 11 for more on microlensing.)

To search for MACHOs, astronomers have monitored the brightness of stars from one of the Milky Way's nearest neighbors, the Large Magellanic Cloud galaxy. To reach Earth, starlight from the Cloud must pass through the halo of the Milky Way galaxy, and MACHOs that reside there should have a measurable effect on that light.

Astronomers have recorded several events in which stars from the Large Magellanic Cloud suddenly brightened and then dimmed again. The number of MACHOs deduced from these observations is nothing much to write home about, however.

Dark matter can be mapped

On much larger scales, scientists are taking advantage of gravitational lensing to map the dark matter in entire galaxies and even clusters of galaxies.

If a cluster happens to lie in the travel path of light emitted by a background galaxy, it will bend and distort that light — *gravitational lensing* — creating multiple images of the background body. A halo of these ghost images forms around the edges of the cluster as seen from Earth.

To create the exact pattern of ghost images that is observed, the intervening cluster must have its mass distributed in a particular way. Because most of the cluster's mass is made of dark matter, this process reveals how the dark matter is concentrated in the cluster.

Dark matter does matter

All methods for detecting and measuring dark matter are indirect, but attempting to understand dark matter is not a trivial pursuit. As the dominant form of matter, dark matter profoundly influences the past, present, and future of the universe.

Antimatter: Opposites Attract

Another type of matter exists that is almost as weird as dark matter. Some say it's weirder. It's called antimatter.

Antimatter was predicted in 1929 by the British physicist Paul Dirac, who combined the theories of quantum mechanics, electromagnetism, and relativity in an elegant set of mathematical equations. (If you want to know more about those theories, you'll have to look them up; this isn't a physics book.) Dirac found that for every subatomic particle, a mirror-image twin should exist, identical in mass but with an opposite electrical charge. Thus the proton has its antiproton, the electron its antielectron.

When a particle and its antiparticle meet, they annihilate each other. Their electric charges cancel out, and their masses are converted into pure energy.

Astronomers have detected antiparticles of the electron and proton in the cosmic rays from deep space. The antielectron is called the *positron* and the antiproton is simply called the *antiproton*. There are also experiments in progress to try and find antihelium in the cosmic rays. Physicists have actually made antiparticles in the laboratory and even entire antiatoms, such as antihydrogen. Doctors use beams of antiparticles to diagnose and treat cancer.

A form of light known as annihilation radiation has been observed by astronomers studying gamma rays from space. Gamma rays are shorter and more energetic than X rays. When an electron and its antiparticle, the positron, meet, they annihilate, releasing gamma rays of known wavelength. These telltale rays have been detected from several places in our galaxy, including a wide region in the direction toward the center of the Milky Way. Annihilation radiation also has been received from some very powerful solar flares.

On the cosmic scale, the big mystery is why the universe contains so many more particles than antiparticles. Experiments are under way to find out why. Presumably, the Big Bang forged equal numbers of both. At least we know we have billions of years to solve the problem before the universe (and us with it!) slips away to whatever fate is in store for it.

This chapter was contributed by Ron Cowen, who covers astronomy and space for *Science News*.

Chapter 16

The Big Bang and the Evolution of the Universe

In This Chapter

▶ Evaluating the evidence for the Big Bang

▶ Understanding inflation and the expansion of the universe

▶ Seeking to know whether the universe is accelerating

▶ Examining the cosmic microwave background

▶ Measuring the Hubble constant and the age of the universe

*O*nce upon a time, about 12 billion years ago, the universe as we know it did not exist. There was no matter — not even an atom. There was no light — not even a photon. Space had yet to be created, and the cosmic clock had yet to start ticking.

Then, perhaps in an instant, the universe took form as a tiny dense speck filled with light. In a minuscule fraction of a second, all the matter and energy in the cosmos came into being. Much smaller than an atom, the infant universe was searingly hot, a fireball that began mushrooming in size and cooling at a furious rate.

This picture of the birth of the universe has come to be known as the *Big Bang* theory.

The Big Bang wasn't so much like a firecracker exploding into existing space, but the rapid expansion of space itself.

During the first trillion-trillion-trillionth of a second, the universe grew more than a trillion-trillion-trillion times bigger. From a smooth mixture of subatomic particles and radiation present then, there later arose the collection of galaxies, galaxy clusters, and superclusters present in the universe today. It's mind boggling to think that the largest structures in the universe, congregations of galaxies that stretch hundreds of millions of light-years across the sky, began as subatomic fluctuations in the energy of the infant cosmos. But that's what scientists currently believe about the way the universe took shape.

Evidence for the Big Bang

Why believe that the universe began with a bang?

Astronomers cite three very different lines of reasoning that make a compelling case for the theory:

- ✔ **The discovery that the universe is expanding.** Perhaps the most convincing evidence for the Big Bang comes from a remarkable discovery made by the American astronomer Edwin Hubble in 1929. Up to that time, most scientists viewed the universe as static — stock still and unchanging. But Hubble found that it's expanding: Groups of galaxies are flying away from each other, like debris flung in all directions from a cosmic explosion (see the section "Hubble's Constant and the Age of the Universe," later in this chapter).

  It stands to reason that if things are flying apart, they were once closer together. Tracing the expansion of the universe back in time, astronomers have deduced that, about 12 billion years ago (give or take a few billion years), the universe was an incredibly hot, dense place in which a tremendous release of energy triggered an enormous explosion.

- ✔ **The discovery of the cosmic microwave background.** In the 1940s, physicist George Gamow realized that the hot Big Bang would produce intense radiation. Colleagues suggested that remnants of this radiation, cooled by the expansion of the universe, might still exist — like the mess left over from a hot party.

  In 1964, Arno Penzias and Robert Wilson of AT&T Bell Laboratories were scanning the sky with a radio receiver when they detected a faint, uniform crackling. What they at first assumed was radio noise turned out to be the faint whisper of radiation left over from the Big Bang. This radiation is a uniform glow permeating all of space at microwave energies. The *cosmic microwave background* has exactly the temperature that astronomers calculated it should have (2.73 kelvins or Celsius degrees above absolute zero, where absolute zero is –273.16°C or –459.69°F) if it had cooled steadily since the Big Bang. For their discovery, Penzias and Wilson shared the 1978 Nobel Prize in physics.

- ✔ **The cosmic abundance of helium.** Astronomers have found that the amount of helium relative to hydrogen in the cosmos is 24 percent. Nuclear reactions inside stars (see Chapter 11) have not gone on long enough to produce this much helium. But the helium is just the amount that the theory predicts would have been forged during the Big Bang.

As successful as the standard Big Bang theory has proved to be in accounting for observations of the cosmos, the theory is but a starting point for exploring the early universe. For example, the theory, despite its name, does not suggest a source for the cosmic dynamite that sparked the Big Bang in the first place.

Inflation: Swell Time in the Universe

The Big Bang theory has other shortcomings besides ignoring the source. It doesn't explain why regions of the universe separated by distances so vast that they could never have communicated — even by a messenger traveling at the speed of light — nonetheless look so similar to each other.

In 1980, physicist Alan Guth invented a theory, which he called *inflation,* that could help explain these puzzles. He suggested that a tiny fraction of a second after the birth of the universe, the universe underwent a tremendous growth spurt. In just 10^{-32} seconds (a hundred-millionth of a trillionth of a trillionth of a second), the universe expanded its girth at a *rate* far greater than at any time in the 12 billion or so years that have elapsed since.

This period of enormous expansion spread tiny regions — which had once been in close contact — out to the far corners of the universe. As a result, the cosmos looks the same, on the large scale, no matter what direction an observer points a telescope. Indeed, inflation expands tiny regions of space into volumes far bigger than astronomers can ever observe. This expansion suggests the intriguing possibility that inflation creates universes far beyond the scope of our own. Instead of a single universe, it's possible that there exists a collection of universes, or a *multiverse.*

Inflation had a second property. This growth spurt captured random, subatomic fluctuations in energy and blew them up to macroscopic proportions. By preserving and amplifying these quantum fluctuations, inflation produced regions with slight variations in density.

Some regions contained more matter and energy, on average, than other regions. This corresponds to cold spots and hot spots in the temperature of the cosmic microwave background (see the preceding section and Figure 16-1). Over time, gravity molded these variations into the spidery network of galaxy clusters and giant voids that fill the universe today.

Figure 16-1:
The light and dark spots in this sky map from the Cosmic Background Explorer (COBE) satellite indicate the hot and cold spots in the cosmic microwave background.

Courtesy of NASA

Something from nothing: Inflation and the vacuum

Ironically, the reservoir of energy that powers inflation comes from nothing: the *vacuum*. According to quantum theory, the vacuum of space is far from empty. It seethes with particles and antiparticles that are constantly being created and destroyed. Tapping into this energy, theorists suggest, provided the explosive energy of the Big Bang and the radiation that was generated along with it.

The vacuum has another bizarre property. It can exert a repulsive gravitational force. Instead of pulling two objects together, *repulsive gravity* draws them farther apart. According to Albert Einstein's Theory of Gravity, the greater the repulsion is, the bigger it can become.

It's this repulsive force that may have led to the brief but powerful era of inflation. Just like economic inflation, cosmic inflation generates a high interest!

Inflation and the shape of the universe

The inflation process, at least is its simplest imagined form, would have imposed another condition on the universe. It would have made the universe flat. Any curvature in the cosmos would have been stretched out by this period of rapid expansion, like a balloon blown to enormous proportions. This is the familiar flat, or Euclidean, geometry of lines and angles drawn on a sheet of paper that you may have learned in high school.

For the universe to be flat, however, it must have contained a very specific density, called the *critical density*. If the density of the universe were greater than the critical value, gravity's pull would be strong enough to reverse the expansion, eventually causing the universe to collapse into what astronomers call the *Big Crunch*.

Such a universe would curve back on itself to form a closed space of finite volume, like the surface of a sphere. A starship traveling in a straight line would eventually find itself back where it had started. Mathematicians call this geometry *positive curvature*.

If the density were less than the critical value, gravity could never overpower the expansion, and the universe would continue to grow forever. Such a universe is said to have *negative curvature*, with a shape akin to a horse's saddle.

Although inflation theory demands that the universe be flat, several types of observations have revealed that the cosmos has only about 40 percent of the density of matter required to keep it flat. When it comes to mass, the cosmic accounting ledger suggests that the universe has come up short.

So if the universe is flat, clumps of matter — be it visible material or invisible, dark matter — can't alone do the trick. There must be some special form of matter or energy (the two are equivalent, according to Einstein) that fills the entire cosmos and makes up the missing 60 percent. Cosmologist Michael Turner of the University of Chicago and the Fermi National Accelerator Laboratory calls this special component *funny energy*.

Funny Energy: Speeding Up Expansion?

Funny energy, if it exists, has a startling consequence. It, too, would exert a repulsive gravitational force. Thus, instead of slowing its expansion since the Big Bang, the universe should be speeding up.

This bizarre notion recently received some unexpected observational support, although the jury still's out on the final results. (For more information on the accelerating universe theory and other concepts in this chapter, visit the University of California, Los Angeles, Frequently Asked Questions in Cosmology site at www.astro.ucla.edu/~wright/cosmology_faq.html).

The new data are based on observations of Type Ia supernovas in distant galaxies. (You can see a picture of this type of supernova in the color section and can read about Type Ia and other supernovas in Chapter 11.)

All supernovas are bright enough to be seen in distant galaxies, but the Ia variety have a special property. Astronomers believe that these explosions all have roughly the same intrinsic brightness, like light bulbs of a known wattage (see the section "Hubble's Constant and the Age of the Universe," later in this chapter).

Because light from a distant galaxy takes hundreds of millions of years to reach Earth, astronomers peering through a telescope at that galaxy may see supernovas that erupted when the cosmos was much younger than it is now. If the universe had been slowing its expansion, there should be less distance between Earth and the faraway galaxy — and a shorter travel time for light — than if the universe had continued to expand at a fixed speed. Thus, in the case of a slower expansion, a supernova from a distant galaxy should look slightly brighter.

But two teams of astronomers found exactly the opposite result: Distant supernovas looked slightly dimmer than expected, as if their home galaxies were farther away than had been calculated. It appears — but is by no means certain — that the universe has revved up its rate of expansion.

This finding is loaded with caveats. Chief among them is that Type Ia super-novas in the distant past may have had a brightness that is different from that of nearby ones — perhaps because their composition was different. If that were the case, astronomers could be fooled into thinking that the dim supernovas mean that the universe is accelerating when all they were seeing were distant supernovas with an intrinsic brightness slightly dimmer than nearby ones.

A new generation of experiments studying the cosmic microwave background (see the preceding section) has begun to weigh in with their own results. A flat universe dictates that the temperature fluctuations — hot and cold spots in the microwave background — have a particular pattern. So far, a slew of balloon-borne and ground-based telescopes suggest that the microwave background does indeed have this pattern.

NASA's Microwave Anisotropy Probe (MAP) satellite is designed to map the microwave background over the entire sky in sharper detail than ever before. (An *anisotropy* is a difference in the physical properties of space, such as tem-perature and density, along one direction from the properties in another direction.) It may provide the most rigorous test yet for inflation, the shape of the universe, and the ultimate fate of the cosmos — whether it will expand forever or gravity will ultimately bring that expansion to a halt and to ulti-mate collapse.

You can follow the progress of MAP by visiting the Web site at map.gsfc.nasa.gov.

Seeds of Galaxy Formation: A Closer Look at the Cosmic Microwave Background

The cosmic microwave background (the faint whisper of radiation left over from the Big Bang) represents a snapshot of the universe when it was about 300,000 years old. Before that time, a fog of electrons pervaded the infant universe, and radiation created in the Big Bang could not stream freely into space. Instead, it was repeatedly absorbed and scattered by these negatively charged particles.

Around the time that the cosmos celebrated its 300,000th birthday, the universe became cool enough for electrons to combine with atomic nuclei. And once these particles combined, the absorbing fog was lifted. Shifted in wavelength by the expansion of the universe, the light from the universe at age 300,000 is today detected as microwaves and far-infrared light.

When the cosmic microwave background was first detected in the 1960s, it appeared to have a perfectly uniform temperature across the sky. There seemed to be no spots that were ever so slightly hotter or colder. That was a puzzle, because such tiny variations in temperature are needed to explain how the universe could have begun as a smooth soup of particles and radiation, but ended up as a lumpy collection of galaxies, stars, and planets.

According to theory, the infant universe was not perfectly smooth. Like lumps in a bowl of porridge, it should have places that are slightly overdense and slightly underdense, with more atoms per cubic inch or fewer atoms per cubic inch, respectively. These represent the tiny seeds around which matter could have started to clump together and galaxies arose. The variations in density should be seen today as tiny fluctuations or anisotropies in the temperature of the cosmic microwave background.

In 1992, NASA's Cosmic Background Explorer satellite, which just three years earlier had measured the temperature of the microwave background to an unprecedented accuracy, achieved what many astronomers consider an even greater triumph: It detected hot and cold spots in the cosmic microwave background.

The variations are indeed minuscule — less than a 10,000th of a degree Kelvin colder or hotter than the average temperature of 2.73 kelvins. Nonetheless, these cosmic ripples are large enough to account for the growth of structure in the universe.

Hubble's Constant and the Age of the Universe

How old is the universe? After years of fractious debate, some astronomers believe that they have pinned down the number — give or take 10 percent or so. By their estimates, the universe is either about 12 billion or about 13.5 billion years old. The former value assumes that the cosmos will expand forever, but at a slower and slower rate; the latter value assumes that some mysterious force is actually speeding up the expansion (see the section Funny Energy: Speeding Up Expansion?" earlier in this chapter).

How fast do galaxies really move?

Cosmic age estimates depend critically on a number that has obsessed astronomers for decades: the Hubble constant, which represents the rate at which the universe is currently expanding. The number dates back to 1929, when astronomer Edwin Hubble found evidence that we live in an expanding universe. In particular, he made the remarkable discovery that every distant galaxy (those beyond the Local Group of Galaxies, which Chapter 12 describes) appears to be racing away from our home galaxy, the Milky Way.

Hubble found that the more remote the galaxy, the faster it is receding. For example, consider two galaxies, one of which lies twice as far from the Milky Way as the other. The galaxy that resides twice as far away appears to move away twice as fast. (According to Albert Einstein's Theory of General Relativity, the galaxies themselves don't move; rather the fabric of space in which they are embedded expands.) This relationship is known as *Hubble's law*.

The constant of proportionality that relates the distance of a galaxy to its recession speed is known as the *Hubble constant,* or H_o. In other words, the speed at which a galaxy is receding is equal to H_o multiplied by the galaxy's distance. H_o thus provides a measure of the rate of expansion of the universe, and by implication, its age.

The Hubble constant is measured in kilometers per second per megaparsec. (One megaparsec is 3.26 million light-years.) After years of study, astronomers using the Hubble Space Telescope (the Earth-orbiting observatory named in honor of Edwin Hubble) recently reported a value for the Hubble constant of 70. That number means that a galaxy about 30 megaparsecs (about 100 million light-years) distant from Earth is speeding away at 2,100 kilometers per second, which is about 1,300 miles per second.

An inconstant constant?

Because the mutual gravitational attraction of galaxies may have slowed the expansion that began with the Big Bang, or some mysterious energy in the cosmos may have recently sped it up, the Hubble constant may in fact not be a constant. The expansion rate could have had a different value in the past. Similarly, the inverse of the Hubble constant (1 divided by H_o), the so-called *Hubble age,* only tells the age of the universe if the expansion rate has been constant since the Big Bang.

Scientists calculate H_o by dividing the the speed at which a galaxy is moving (the *velocity*) by its distance. Obtaining a velocity is easy: Astronomers analyze the specific colors, or wavelengths of light emitted or absorbed by a galaxy. Light from an object speeding away from Earth shifts to redder, or longer, wavelengths; the greater the red shift, the faster the galaxy is receding.

In contrast, measuring distance has proven far trickier.

For starters, in order to accurately measure the expansion rate of the universe, astronomers must gauge the distance to very remote galaxies, those that lie 600 million light-years or more from Earth. At lesser distances, the expansion is partly counteracted by the gravitational tug of galaxies that are relatively near the Milky Way.

However, astronomers have no completely reliable way of directly measuring distances to remote galaxies. Instead, they must resort to a variety of indirect means to size the cosmos. By calibrating the distance to nearby galaxies and working their way outward, step by step, to more distant galaxies, astronomers have pieced together a yardstick for the universe.

How are galaxy distances measured?

Most strategies for measuring distance require some kind of *standard candle,* the cosmic equivalent of a light bulb of known wattage.

For instance, suppose that you believe you know the true brightness or *luminosity* of a particular type of star. Light from a distant source grows dimmer in proportion to the square of the distance. So the faintness of that star in a distant galaxy indicates how far way the galaxy lies.

Yellowish, pulsating stars known as *Cepheid variables* remain one of the most credible standard candles for estimating the distance to relatively nearby galaxies (see Chapter 12). These youthful stars brighten and dim periodically.

In 1912, Henrietta Leavitt of Harvard College Observatory detected that the rapidity with which Cepheids change their brightness is directly linked to their true luminosity. The longer the period, the greater the luminosity.

Type Ia supernovas (see the earlier section, "Funny Energy: Speeding Up Expansion?" and Chapter 11) are another type of standard candle. Because supernovas are much brighter than Cepheids, they can be seen in galaxies that are much more distant. Recent calculations of the Hubble constant employ both of these candles as well as two other types of calibrators.

These methods are still rather rough, so although we know for sure that the universe is expanding, its exact speed and how that speed has changed over the eons of cosmic time are not well known. The equivalent of a cosmic cop may be standing somewhere with a radar gun that can track the speed of expansion, but it's pretty hard to look over his shoulder and see what that speed is!

This chapter was contributed by Ron Cowen, who covers astronomy and space research for *Science News*.

Part V
The Part of Tens

The 5th Wave   By Rich Tennant

In this part . . .

Did you ever find yourself at a social gathering, desperately trying to think of something unique and interesting to say? You searched your brain for some crowd-grabbing insight that would make everyone in the room take notice of your remarkable intelligence. Well, read this part, and you'll be ready for the next lapse in conversation. I offer you ten strange facts about space that are guaranteed to garner interest. And then I fill you in about ten major mistakes that people in general, and the media in particular, have made, and keep making, when the topic is astronomy.

Chapter 17

Ten Strange Facts about Astronomy and Space

In This Chapter

▶ Getting the truth about comet tails, Mars rocks, meteorites in your hair, and the Big Bang on black-and-white TV

▶ Finding out why Pluto's discovery was an accident, sunspots aren't dark, and rain never hits the ground on Venus

▶ Exploring tidal myths, exploding stars, and Earth's uniqueness

Here are some of my favorite facts about astronomy and, especially, Earth and its solar system. With this information under your belt, you may be ready to handle the astronomy questions on TV shows such as *Who Wants To Be A Millionaire?*

A Comet's Tail Often Leads the Way, Instead of Trailing Behind

A comet tail is not like a horse's tail, which is always behind as the horse gallops ahead. A comet tail always points away from the Sun. So when a comet approaches the Sun, its tail or tails stream behind it, but when the comet is heading back out into the solar system, the tail leads the way. (See Chapter 4 for more information about comets.)

Rocks from Mars Are All Over Earth

About a dozen meteorites have been found on Earth that are fragments from the crust of Mars, blasted from that planet by the impacts of much larger objects, perhaps from the asteroid belt. But the Mars rocks that have been found are only the ones that were recognized by meteorite hunters or actually seen to fall by witnesses. Statistically, many more must have fallen into the ocean or landed in out-of-the-way places where they have never been discovered. (See Chapter 6 to find out about Mars.)

There Are Tiny Meteorites in Your Hair

Micrometeorites, tiny particles from space that are visible only through microscopes, are constantly raining down on Earth. Some are falling on you whenever you go outdoors. But without the most advanced laboratory equipment and analysis techniques, you can't detect them. They are lost in the great mass of pollen, smog particles, household dust, and probably (I'm sorry to say) dandruff that reside on top of your head.

You May Have Seen the Big Bang on an Old TV

Some of the "snow" (a pattern of interference that looks like little white spots or streaks on old black-and-white TV sets) is actually radio waves received from the microwave background radiation, a glow from the early universe in the aftermath of the Big Bang (see Chapter 16). When this radiation was actually discovered at the Bell Telephone Laboratories, many possible causes of the unexpected "noise" in the radio receiver were studied. Scientists even investigated pigeon droppings as a possible cause.

Pluto Was Discovered from the Predictions of a False Theory

Percival Lowell predicted the existence and approximate location of Pluto. When Clyde Tombaugh surveyed the region, he discovered the planet. But now we know that Lowell's theory, which inferred the existence of Pluto from its gravitational effects on the motion of Uranus, was wrong. In fact, Pluto's mass is very low and incapable of producing the "observed" effects. Furthermore, these effects were just errors in measuring the motion of Uranus. (Not enough information was available about Neptune's motion to study it for clues to the existence of Pluto.) The discovery of Pluto took hard work, but it was just plain luck. (See Chapter 9 to find out more about Pluto.)

Sunspots Are Not Dark

Everyone "knows" that sunspots are those "dark" spots on the Sun. But in reality, sunspots are simply places where the hot solar gas is slightly cooler than its surroundings (see Chapter 10 for more explanation). The spots look dark compared to their hotter surroundings, but if all you saw was the sunspot, it would look very bright.

On Venus, the Rain Never Falls on the Plain

In fact, the constant rain on Venus never falls on anything. It evaporates before it hits the ground, and the rain is pure acid. (See Chapter 6.)

The Ocean Tides Facing the Moon Are No Higher Than Tides on the Other Side of Earth

The fact defies common sense, but not physics and mathematical analysis: The tides that the Moon raises in the ocean on Earth's hemisphere that is facing the Moon are no higher than the tides that occur at the same time on the opposite side of Earth (see Chapter 5 for more about the Moon). (And the same goes for the smaller ocean tides that are raised by the Sun.)

A Star in Plain View May Have Erupted in an Enormous Supernova Explosion, but No One Can Tell

Eta Carinae is one of the most massive and fiercely shining stars in our galaxy, and is expected to produce a powerful supernova explosion at any time, if not already. But because it takes light 9,000 years to travel from

Eta Carinae to Earth, an explosion that occurred less than 9,000 years ago would not be visible to us yet. (See Chapter 11 to discover more about the life cycles of stars.)

Earth Is Made of Rare and Unusual Matter

The great majority of all the matter in the universe is so-called *dark matter,* invisible stuff that astronomers have not yet identified (see Chapter 15). And even among the ordinary or visible matter, most is in the form of plasma (hot electrified gas that makes up normal stars such as Earth's Sun) or degenerate matter (in which atoms or even the nuclei within the atoms are crushed together to unimaginable density, as found in white dwarf and neutron stars, which I describe in Chapter 11). There's no dark matter, no degenerate matter, and hardly any plasma on Earth. Compared to the great bulk of the universe, our planet and we are the aliens. (See Chapter 5 for more about Earth's unique properties.)

Chapter 18

Ten Common Errors about Astronomy and Space

In This Chapter
▶ Popular misconceptions about astronomy
▶ Mistakes commonly made by the news and entertainment media

*I*n daily life — reading the newspaper, watching the evening news, or talking to friends — you will run across several frequently repeated mistakes on astronomy. In this chapter, I explain some of these common misconceptions.

If You Were in the Asteroid Belt, You Would See Asteroids All Around You

In just about any movie about space travel, there's a scene in which the intrepid pilot skillfully steers the spaceship past hundreds of asteroids that hurtle past in every direction, sometimes coming five at a time. Moviemakers just don't understand the vastness of the solar system. If you were standing on an asteroid smack dab in the middle of the main asteroid belt, between Mars and Jupiter, you would be lucky to see more than one or two other asteroids, if any, with the naked eye. (See Chapter 7 for more information about asteroids.)

Nuking a "Killer Asteroid" on a Collision Course for Our Planet Will Save Earth

There are many common errors about asteroids, and the recent spate of doomsday movies and media reports on "killer asteroids" have provided ample but unfortunate opportunity to reinforce these misunderstandings among the public.

If an asteroid is on a collision course with Earth, blowing it up with an H-bomb will only make a lot of smaller and collectively-just-as-dangerous rocks, still heading for our planet. Before *Armageddon* came out, I was a Bruce Willis fan, but now I'm only a demi-admirer.

Asteroids Are Round, Like Little Planets

A few of the largest asteroids *are* round, but the great majority are irregular blocks of stone or iron. Many are shaped like peanuts or potatoes, and are pitted with craters. (Check out Chapter 7 to find out more about asteroids.)

The Big Bang Is Dead

When an astronomer reports a finding that doesn't fit the current understanding of cosmology, the media often bray that "the Big Bang is dead." (See Chapter 16 for an explanation of the Big Bang.) But astronomers are simply finding differences between the observed expansion of the universe and specific mathematical descriptions of it. The competing theories — including one that fits the newly reported data — are also consistent with the Big Bang; they just differ in the details.

A Meteorite That Just Fell to the Ground Is "Still Hot"

Actually, freshly fallen meteorites are cold; an icy frost sometimes forms (from contact with moisture in the air) on a frigid stone that has just landed. When an eyewitness says that he saw a meteorite fall to the ground and he claims that he burned his fingers on the rock, the account is probably a hoax. (See Chapter 4 for more information about meteorites.)

Summer Happens When the Earth Is Closest to the Sun

The belief that summer is when Earth is closest to the Sun is about the most common error of them all, but common sense tells you that it's false. After all, winter occurs in Australia when the United States is experiencing summer. But on any given day, Australia is the same distance from the Sun as the United States. (See Chapter 5 for more explanation.)

"The Light from That Star Took 1,000 Light-Years to Reach Earth"

Many people mistake the light-year for a unit of time, like a day, month, or an ordinary year. But a light-year is a unit of distance, equal to the length that light travels in a vacuum over a period of one year. (See Chapter 1.)

When the Distance of a Galaxy Is Reported as, Say, "Two Billion Light-Years," That's a Fact

Astronomers' knowledge of the distances of very distant galaxies (those hundreds of millions of light-years away, or more) is so inaccurate that for professional purposes, we never publish the estimated distances. Because the media demand specifics, astronomers make statements like "if such-and-such version of the Big Bang is the correct theory, the distance of this galaxy is two billion light-years." But reporters simplify their accounts and omit those weasel words. Until recently, the reported values could be in error by as much as 200 percent.

The "Morning Star" Is a Star

The Morning Star is never a star, it's always a planet. And sometimes two Morning Stars appear at once, such as Mercury and Venus (see Chapter 6). The same idea applies to the Evening Star. It's a planet, and there may be more than one. "Shooting stars" and "falling stars" are misnomers, too. They aren't stars; they are meteors, the flashes of light caused by small meteoroids falling through Earth's atmosphere.

The Sun Is an Average Star

You can find the statement that the Sun is an average star repeated by journalists and even published in books written for the general public by astronomers who should know better. In fact, the vast majority of all stars are smaller, dimmer, cooler, and less massive than our Sun (see Chapter 10). Be proud of the Sun — it's like a child from Lake Wobegon.

The Hubble Telescope Gets Up Close and Personal

The Hubble Space Telescope does *not* get those beautiful pictures by cruising through space until it's alongside those nebulae, star clusters, and galaxies that appear in the pictures. The telescope stays in close orbit around Earth, and it just takes great photos!

Part VI
Appendixes

The 5th Wave   By Rich Tennant

"Paul, turn off your flashlight. There's a real interesting star cluster I'm trying to get a picture of."

In this part . . .

The appendixes in this part offer information that will enhance your skywatching experiences for years to come. The first appendix offers maps to help you find interesting stars. The second one gives you tables showing the approximate locations — at any time from the year 2000 through 2004 — of the four bright planets that are the most easy to spot: Venus, Mars, Jupiter, and Saturn. Finally, I include simple definitions for some astronomy terms that you'll use as you enjoy your sky-watching hobby.

Appendix A

Finding the Planets: 2000 to 2004

• •

*T*he tables on the following pages give, for the years 2000 through 2004, approximate locations of the four bright planets that are the most commonly observed: Venus, Mars, Jupiter, and Saturn. These planets are usually easy to spot with the unaided eye, and once found, can be tracked for several consecutive months. For each year, I include separate tables for the twilight periods: dawn and dusk. These times are the most convenient for most people to observe the sky. In each table, I indicate which direction to look for the planets. The tables are most accurate for middle northern latitudes.

If you follow the Moon daily at dusk or dawn, you'll very often spot it near one of the five bright planets (Mercury, Venus, Mars, Jupiter, or Saturn), or near one of the several bright stars or noteworthy patterns in the Zodiac: the Pleiades, Hyades, or Aldebaran in *Taurus;* Pollux and Castor in *Gemini;* Regulus in *Leo;* Spica in *Virgo;* Antares in *Scorpius;* and the Teapot in *Sagittarius.* (See Chapter 3 for more information about the Zodiac.)

As you follow the planets over days, weeks, or months, you'll notice that they drift past one another, as well as past the same Zodiacal stars the Moon does. Planet-watching is an activity that can provide enjoyment for a lifetime!

A highly recommended resource for planet-gazers is the *Sky Calendar* from Abrams Planetarium. For information, write to Sky Calendar, Abrams Planetarium, Michigan State University, East Lansing, MI 48824. These charts were supplied by Robert Victor, who has since retired from Abrams Planetarium.

2000

Table A-1	Planets at Dusk (About 45 Minutes after Sunset)				
Month	*Venus*	*Mars*	*Jupiter*	*Saturn*	*Planet Happenings*
January	—	SW	S	SE	—
February	—	WSW	SW	SW	During the first 3 weeks, four bright planets are visible simultaneously during Mercury's fine appearance low in the WSW at dusk.
March	—	W	W	WSW	—
April	—	Low WNW	Low WNW	Low WNW	On April 5, Mars passes Jupiter; 6 through 17, all three bright outer planets fit within the same binocular field; 15, Mars passes Saturn, and a compact gathering (only 5 degrees long) of three outer planets occurs.
May	—	Setting WNW	—	—	May 18, emerging Mercury is guide to faint departing Mars.
June	—	—	—	—	—
July	—	—	—	—	—
August	Setting W	—	—	—	—
September	Low WSW	—	—	—	September 26 to 28, Mercury appears to the lower right of Venus, in the same binocular field.
October	Low SW	—	—	—	—
November	SW	—	Rising ENE	Low ENE	November 19, Saturn is visible all night; 27, Jupiter is visible all night.
December	SW	—	E	E	—

Table A-2   Planets at Dawn (About 45 Minutes before Sunrise)

Month	Venus	Mars	Jupiter	Saturn	Planet Happenings
January	SE	—	—	—	—
February	Low SE	—	—	—	—
March	Rising ESE	—	—	—	March 16, Mercury is above Venus.
April	—	—	—	—	—
May	—	—	—	—	May 17, a very close pairing of Venus and Jupiter occurs, but the sight will be lost in the Sun's glare; 18, a compact gathering of Venus, Jupiter, and Saturn will also be lost in the Sun's glare; 28, Jupiter passes Saturn, visible from the tropics and the southern latitudes.
June	—	—	Low ENE	Low ENE	Emerging from the solar glare, Jupiter and Saturn are 1.5 degrees apart on June 6, and 2 degrees apart on June 13. For the next pairing of these slow-moving giant planets, we must wait until December 21, 2020.
July	—	—	E	E	—
August	—	Low ENE	ESE	SE	August 10, departing Mercury is guide to faint emerging Mars.
September	—	E	S	SSW	—
October	—	ESE	WSW	WSW	—
November	—	SE	W	Low WNW	—
December	—	SSE	Setting WNW	—	—

2001

Table A-3	Planets at Dusk (About 45 Minutes after Sunset)				
Month	**Venus**	**Mars**	**Jupiter**	**Saturn**	**Planet Happenings**
January	SW	—	ESE	ESE	Middle to late January, Venus appears half full through a telescope; middle January to early February, Mercury makes a favorable appearance in the low WSW, at the far lower right of Venus. Four planets are visible simultaneously.
February	WSW	—	S	SSW	February and March, Venus shows its crescent phase and is best observed with binoculars or a telescope at sunset.
March	W	—	WSW	WSW	March, Venus shows greatest brilliancy.
April	—	—	W	W	—
May	—	—	Low WNW	Setting WNW	During most of May, Mercury makes a very favorable appearance. It passes Saturn on May 6, and Jupiter on May 15.
June	—	Low SE	—	—	Throughout June, Mars is unusually bright. It's visible all night on June 13 and closest to Earth on June 21.
July	—	SSE	—	—	—
August	—	S	—	—	—
September	—	S	—	—	—
October	—	S	—	—	—
November	—	S	—	Rising ENE	—
December	—	S	Rising ENE	ENE	December 3, Saturn is visible all night; and on 31, Jupiter is visible all night.

Table A-4   Planets at Dawn (About 45 Minutes before Sunrise)

Month	Venus	Mars	Jupiter	Saturn	Planet Happenings
January	—	S	—	—	—
February	—	S	—	—	—
March	—	S	—	—	—
April	Low E	S	—	—	Throughout April and May, Venus shows its crescent phase, and is best observed with binoculars or telescope at sunrise.
May	E	SSW	—	—	Venus is at greatest brilliancy.
June	E	low SW	—	Rising ENE	Early June, Venus appears half full viewed through a telescope.
July	E	—	Low ENE	E	After Mercury and Jupiter emerge in the ENE in early July, four planets are visible simultaneously; 13, Mercury passes Jupiter, and they fit within a 5-degree field from 8 to 16; 15, Venus passes Saturn. Venus, Saturn, and the bright star Aldebaran fit within a 5-degree field from July 11 to 17.
August	E	—	E	ESE	August 5 and 6, Venus passes Jupiter.
September	E	—	ESE	SSE	
October	E	—	S	WSW	From October 28 to November 7, Mercury and Venus are within 1 degree of each other.
November	Low ESE	—	WSW	W	From late October to mid-November, four planets are visible simultaneously.
December	Rising ESE	—	W	Setting WNW	—

2002

Table A-5   Planets at Dusk (About 45 Minutes after Sunset)

Month	Venus	Mars	Jupiter	Saturn	Planet Happenings
January	—	SSW	E	ESE	January 1 to 20, four planets are visible simultaneously at dusk during Mercury's favorable appearance low in the WSW.
February	Setting in W	WSW	ESE	SSE	After Venus emerges late in the month, at least four planets are visible at dusk until late May.
March	Low W	W	S	WSW	—
April	WNW	W	WSW	W	In late April and early May, during Mercury's fine appearance low in the WNW, all five bright planets are visible simultaneously.
May	WNW	WNW	W	Low WNW	May 2, Mars passes Saturn; 5 and 6, Venus, Mars, and Saturn form a compact triangle; 6, Venus passes Saturn; 10, Venus passes Mars.
June	WNW	Setting WNW	Low WNW	—	June 3, Venus passes Jupiter.
July	W	—	—	—	—
August	WSW	—	—	—	Mid-August, Venus appears half full.
September	Low WSW	—	—	—	September and early October, Venus appears as a crescent, and is best observed at sunset. It reaches greatest brilliancy in September.
October	Setting SW	—	—	—	—
November	—	—	—	—	—
December	—	—	—	Low ENE	December 17, Saturn is visible all night.

Table A-6   Planets at Dawn (About 45 Minutes before Sunrise)

Month	Venus	Mars	Jupiter	Saturn	Planet Happenings
January	—	—	Setting WNW	—	—
February	—	—	—	—	—
March	—	—	—	—	—
April	—	—	—	—	
May	—	—	—	—	—
June	—	—	—	Rising ENE	—
July	—	—	—	ENE	July 2, Mercury passes Saturn.
August	—	—	Low ENE	E	—
September	—	Rising E	E	SE	—
October	—	E	ESE	SW	Most of October offers a fine view of Mercury low in the E to ESE; 9 to 11, Mercury passes near Mars. Four planets are visible simultaneously during Mercury's appearance.
November	ESE	ESE	S	W	Four planets are visible simultaneously when Venus emerges. In November and December, Venus appears as a crescent, and is best observed at sunrise.
December	SE	SE	WSW	WNW	In early December, Venus is at greatest brilliancy, near Mars from December 4 to 8.

2003

Table A-7				Planets at Dusk (About 45 Minutes after Sunset)	
Month	*Venus*	*Mars*	*Jupiter*	*Saturn*	*Planet Happenings*
January	—	—	Rising ENE	E	—
February	—	—	E	SE	February 2, Jupiter is visible all night.
March	—	—	ESE	SW	Saturn's rings are at greatest inclination (27 degrees) from edge-on.
April	—	—	S	W	End of March until late April, Mercury makes a very favorable appearance low in the W to WNW.
May	—	—	WSW	WNW	—
June	—	—	W	Setting WNW	—
July	—	—	Low WNW	—	July 25, Mercury passes Jupiter.
August	—	Rising ESE	—	—	August 27, Mars is nearest to Earth, unusually brilliant, and visible nearly all night; this is the closest approach of Mars in over 2,000 years, and the nearest until the year 2287.
September	—	SE	—	—	—
October	Setting WSW	SE	—	—	—
November	Low SW	SE	—	—	—
December	SW	SSE	—	Rising ENE	December 4 and 5, Mercury is to the lower right of Venus; 31, Saturn is visible all night.

Table A-8 Planets at Dawn (About 45 Minutes before Sunrise)

Month	Venus	Mars	Jupiter	Saturn	Planet Happenings
January	SE	SSE	W	—	Middle and late January, Venus appears half full through a telescope.
February	SE	SSE	Setting WNW	—	During Mercury's favorable appearance in the ESE from late January to mid-February, four planets are visible simultaneously.
March	Low ESE	SSE	—	—	—
April	Low E	SSE	—	—	—
May	Low E	SSE	—	—	May 26 and 27, Mercury passes Venus, visible in binoculars, from the southern United States.
June	Low ENE	SSE	—	—	June 21, Use binoculars to see Mercury pass Venus.
July	Rising ENE	S	—	Low ENE	July 8, Venus passes Saturn, and is visible in binoculars, from the southern United States.
August	—	SW	—	E	—
September	—	Setting WSW	Low E	ESE	September 21 and 22, Mercury is low in the E, to the lower left of Jupiter.
October	—	—	ESE	S	Mercury makes a favorable appearance during late September and early October.
November	—	—	SE	WSW	—
December	—	—	SSW	W	—

2004

Table A-9		Planets at Dusk (About 45 Minutes after Sunset)			
Month	*Venus*	*Mars*	*Jupiter*	*Saturn*	*Planet Happenings*
January	SW	S	—	E	—
February	WSW	SW	Rising E	ESE	Late February through May, at least four planets are visible at dusk.
March	W	WSW	E	S	March 4, Jupiter is visible all night. Mid-March to early April provides a fine view of Mercury in the W to WNW; all five bright planets visible!
April	W	W	SE	SSW	April until the start of June, Venus shows its crescent through binoculars or telescope, best observed around sunset. April 23 to 25, Venus not far lower right of Mars.
May	WNW	WNW	SSW	W	Early May, Venus reaches greatest brilliancy; May 24, Mars passes Saturn.
June	—	WNW	WSW	Setting WNW	—
July	—	Setting WNW	W	—	July 10, Mercury is closely above faint Mars.
August	—	—	Low W	—	—
September	—	—	—	—	—
October	—	—	—	—	—
November	—	—	—	—	—
December	—	—	—	—	—

Table A-10   Planets at Dawn (About 45 Minutes before Sunrise)

Month	Venus	Mars	Jupiter	Saturn	Planet Happenings
January	—	—	WSW	Setting WNW	—
February	—	—	W	—	—
March	—	—	Low W	—	—
April	—	—	—	—	—
May	—	—	—	—	—
June	Rising ENE	—	—	—	Late June to early August, Venus appears as a crescent through binoculars or telescope, best observed at sunrise.
July	E	—	—	Rising ENE	Mid-July, Venus is at greatest brilliancy.
August	E	—	—	ENE	In mid-August, Venus appears half full. August 31 and September 1, Venus passes Saturn.
September	E	—	—	E	Most of September, Mercury is in fine view low in the E at dawn, to the far lower left of Venus.
October	ESE	Rising E	Low E	SSE	Beginning late in October after Jupiter and Mars have emerged, at least four planets can be observed simultaneously at dawn for the rest of the year.
November	ESE	ESE	SE	SW	November 4 and 5, Venus passes Jupiter.
December	SE	SE	SSE	W	December 5 and 6, Venus passes Mars; when Mercury emerges after mid-month, all five bright planets are visible. From December 26 well into January 2005, Mercury lingers very near Venus.

Appendix B

Star Maps

The following pages contain eight star maps, four each of the Northern and Southern Hemispheres, to help start you on your starry way.

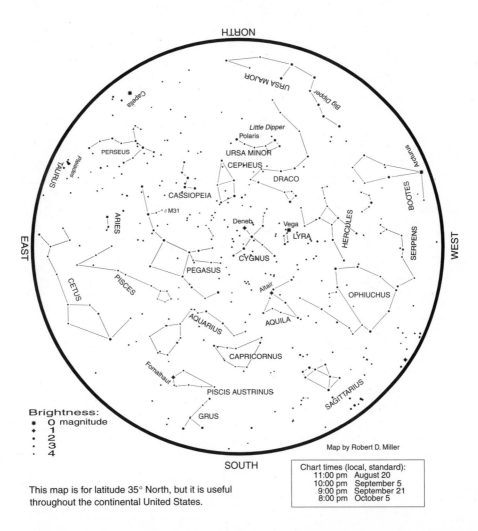

Brightness:
* 0 magnitude
* 1
* 2
* 3
* 4

Map by Robert D. Miller

This map is for latitude 35° North, but it is useful throughout the continental United States.

Chart times (local, standard):
11:00 pm   August 20
10:00 pm   September 5
9:00 pm   September 21
8:00 pm   October 5

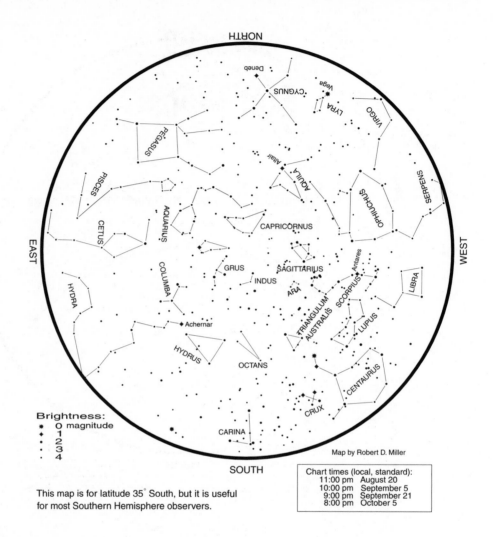

Brightness:
* ✱ 0 magnitude
* ◆ 1
* • 2
* · 3
* · 4

Map by Robert D. Miller

Chart times (local, standard):
11:00 pm   August 20
10:00 pm   September 5
9:00 pm   September 21
8:00 pm   October 5

This map is for latitude 35° South, but it is useful for most Southern Hemisphere observers.

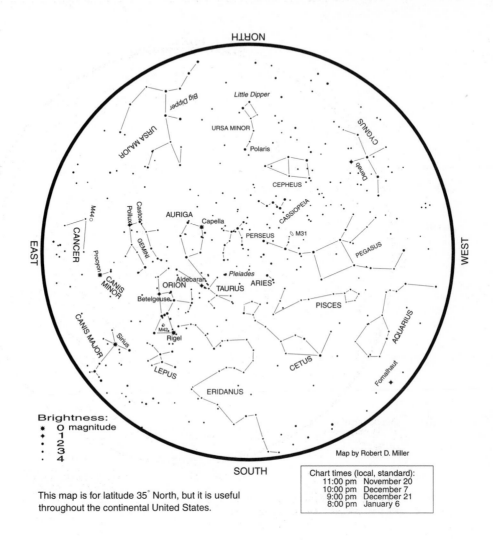

Map by Robert D. Miller

Brightness:
* ✳ 0 magnitude
* ✦ 1
* • 2
* · 3
* · 4

This map is for latitude 35° North, but it is useful throughout the continental United States.

Chart times (local, standard):
11:00 pm   November 20
10:00 pm   December 7
9:00 pm   December 21
8:00 pm   January 6

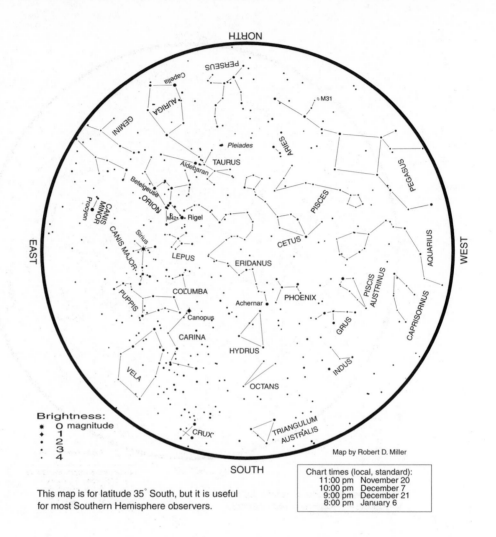

Brightness:
✳   0 magnitude
✦   1
•   2
·   3
·   4

This map is for latitude 35° South, but it is useful
for most Southern Hemisphere observers.

Map by Robert D. Miller

Chart times (local, standard):
   11:00 pm   November 20
   10:00 pm   December 7
    9:00 pm   December 21
    8:00 pm   January 6

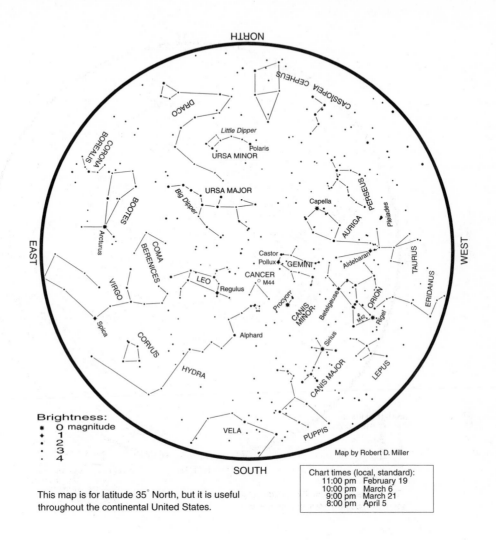

Map by Robert D. Miller

Brightness:
* ✳ 0 magnitude
* ✦ 1
* • 2
* · 3
* · 4

This map is for latitude 35° North, but it is useful throughout the continental United States.

Chart times (local, standard):
11:00 pm February 19
10:00 pm March 6
9:00 pm March 21
8:00 pm April 5

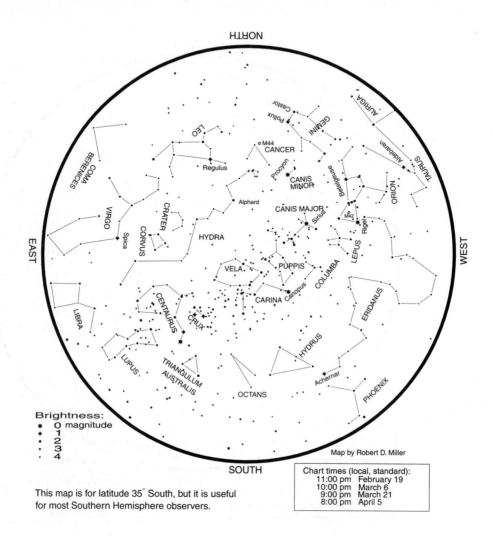

Map by Robert D. Miller

Brightness:
* ✳ 0 magnitude
* ✦ 1
* • 2
* • 3
* · 4

This map is for latitude 35° South, but it is useful for most Southern Hemisphere observers.

Chart times (local, standard):
  11:00 pm   February 19
  10:00 pm   March 6
  9:00 pm   March 21
  8:00 pm   April 5

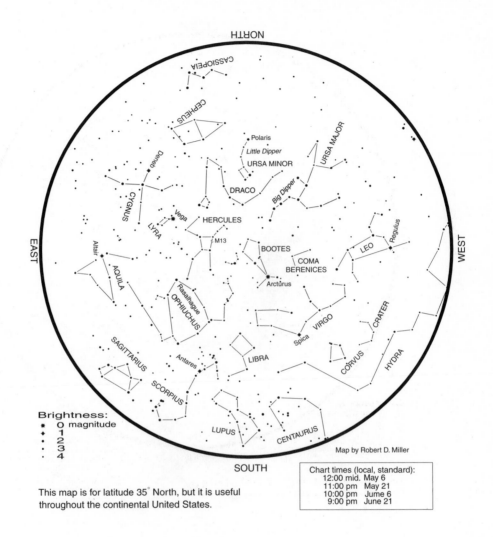

Brightness:
* 0 magnitude
* 1
* 2
* 3
* 4

Map by Robert D. Miller

Chart times (local, standard):
12:00 mid. May 6
11:00 pm May 21
10:00 pm Jume 6
9:00 pm June 21

This map is for latitude 35° North, but it is useful throughout the continental United States.

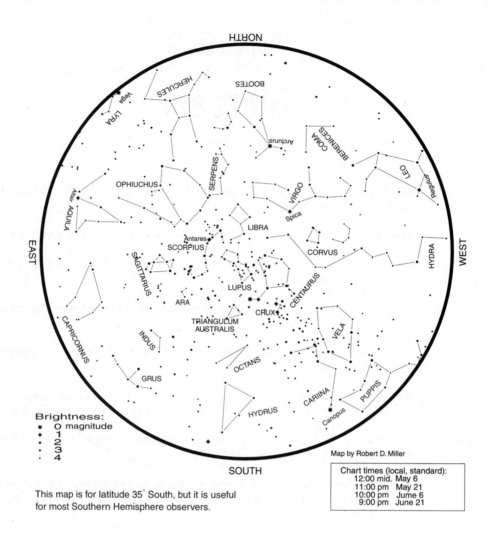

Brightness:
* 0 magnitude
* 1
* 2
* 3
* 4

Map by Robert D. Miller

Chart times (local, standard):
12:00 mid. May 6
11:00 pm May 21
10:00 pm Jume 6
9:00 pm June 21

This map is for latitude 35° South, but it is useful
for most Southern Hemisphere observers.

Appendix C

Glossary

asterism A named pattern of stars, such as the Big Dipper, that is not one of the 88 official constellations

asteroid One of many small, rocky, and/or metallic bodies orbiting the Sun

binary star Two stars orbiting around a common center of mass in space; also called a **binary system**

black hole An object with gravity so strong that nothing inside can escape, not even a ray of light

comet One of many small bodies made of icy and dusty matter orbiting the Sun

crater A round depression on the surface of a planet, moon, or asteroid, due to the impact of a falling body, a volcanic eruption, or the collapse of an area

dark matter Unknown substance in space whose presence is detected by its gravitational effect on celestial objects

Doppler effect The process by which light or sound is altered in perceived frequency or wavelength by the motion of its source with respect to the observer

double star Two stars that look very close to each other on the sky and may be physically associated (a **binary star**), or may be unrelated to each other and at different distances from Earth

ecliptic The apparent path of the Sun across the background of the constellations

galaxy A huge system of billions of stars, sometimes with vast amounts of gas and dust

meteor The flash of light caused by the fall of a meteoroid through Earth's atmosphere; often incorrectly used to mean the meteoroid itself

meteorite A meteoroid that landed on Earth

meteoroid A rock in space, composed of stone and/or metal; probably a chip from an asteroid

nebula A cloud of gas and dust in space that may emit, reflect, and/or absorb light

neutron star An object only tens of miles across, but greater in mass than the Sun (all pulsars are neutron stars, but not all neutron stars are pulsars)

occultation The process by which one celestial body passes in front of another, blocking it from the view of an observer

orbit The path followed by a celestial body or a spacecraft

planet A large, round object that formed in a flattened cloud around a star and which — unlike a star — doesn't generate energy by nuclear reactions

planetary nebula Glowing, expanding gas cloud that was expelled in the death throes of a sunlike star

pulsar A fast-spinning, tiny, and immensely dense object that emits light, radio waves, and/or X-rays in one or more beams like the beam from a lighthouse

quasar A small, extremely bright object at the center of a distant galaxy, thought to represent the emission of much energy from the surroundings of a giant black hole

red giant A large, very bright star with a low surface temperature; a late stage in the life of a sunlike star

redshift An increase in the wavelength of light or sound, often due to the Doppler Effect

rotation The spinning of an object around an axis that passes through it

seeing A measure of the steadiness of the air at a place where astronomical observations are being made (when the seeing is good, the images viewed through telescopes are sharper)

star A large mass of hot gas held together by its own gravity and fueled by nuclear reactions

star cluster A group of stars held together by their mutual gravitational attraction and which formed together at about the same time (types include **globular clusters** and **open clusters**)

supernova An immense explosion that disrupts an entire star and that may form a black hole or a neutron star

terminator The line separating the illuminated and unilluminated parts of a body that shines by reflected light of the Sun, such as a planet, a moon, or an asteroid

transit The movement of a smaller object, such as Mercury, in front of a larger object, such as the Sun

variable star A star that changes perceptibly in brightness

white dwarf A small, dense object shining from stored heat and thus fading away; the final stage in the life of a sunlike star

zenith The point on the sky that is directly above the observer

SkyMeasures

arc minutes/arc seconds Units of measurement on the sky. A full circle around the sky consists of 360 degrees, each divided into 60 arc minutes; each arc minute is divided into 60 arc seconds.

astronomical unit (A.U.) A measure of distance in space, equal to the average distance between Earth and the Sun, about 93 million miles.

declination On the sky, the coordinate that corresponds to latitude on Earth, and is measured in degrees north or south of the Celestial Equator.

light-year The distance light travels through space in one year; about 5.9 trillion miles.

magnitude A measure of the relative brightness of stars, with smaller magnitudes corresponding to brighter stars. For example, a first magnitude star is 100 times brighter than a sixth magnitude star.

right ascension On the sky, a coordinate that corresponds to longitude on Earth, and is measured eastward from the vernal equinox, (a point in the sky where the celestial equator crosses the ecliptic and where the Sun is located on the first day of spring in the Northern Hemisphere).

Index

• A •

AAVSO (American Association of Variable Star Observers), 194
abbreviations of star/constellation names, 16, 17–20
absolute magnitude, 48
accretion disk, 221
Achernar, 45
Acrux, 45
active galactic nuclei (AGN), 224–227
age of sky objects, 87–88, 146–147, 245, 258
AGNs (active galactic nuclei), 224–227
albedo map, 112
Aldebaran, 45
Algol, 191
Alpha Canis Majoris. *See* Sirius
Alpha Centauri, 192
Alpha Lyrae, 193
Alpha Orionis, 193
alphabet, Greek, 15
ALPO (Association of Lunar and Planetary Observers), 114
alt-azimuth telescope mounts, 55
Altair, 45
altitude, 55, 66
AM Herculis systems, 189
AMANDA (neutrino observatory), 154
American Association of Variable Star Observers (AAVSO), 194
ammonium hydrosulfide, 125
Andromeda Galaxy, 197, 211
anisotropy, 256
annihilation radiation, 249
annular eclipses, 164
Antares, 45, 110
antimatter, 249
antipode of Caloris, 100
apparent magnitude, 48
apparent size, 109
Arcturus, 45
Arecibo radio telescope, 234
artificial satellites, 73–75, 102
ashen light, 110
"Ask a NASA Scientist," 33
aspects of inferior planet, 108
aspects of superior planet, 107
Association of Lunar and Planetary Observers (ALPO), 114
associations, 196, 199–202
asterisms, 293
Asteroid Belt, 118

asteroidal meteoroids, 60, 118
asteroidal occultations, 123
asteroids. *See also* Near Earth Objects (NEOs)
    about, 117–118
    "Big Four," 122
    defined, 293
    errors about, 267–268
    motion of, 118
    naming conventions, 74
    nudging near-earth, 120
    protecting Earth from, 121
    viewing, 122–123
astrology, 11
Astronomical Calendar, 113
Astronomical League, 32
Astronomical Society of the Pacific, 32, 94
Astronomical Units (A.U.s), 24
astronomy
    defined, 10–11
    with family, 26
    plan for starting hobby, 58
    practicing, 9
    what it isn't, 11
astronomy clubs, 32–33
Astronomy magazine, 33, 34, 130
Astronomy Mall, 33
atmosphere, 81, 96, 100, 115
atmospheric turbulence, 51. *See also* seeing
atoms, 87, 185, 206–207, 247
auroras, 98
A.U.s (Astronomical Units), 24
autumnal equinox, 87
averted vision, 108
axions, 246
axis, Earth's, 85–86
azimuth, 55

• B •

bad seeing, 51
Baily's Beads, 166
Baltis Vallis, 101
bare black holes, 221. *See also* black holes
barred spiral galaxies, 208–209. *See also* galaxies
Baryonic dark matter, 246
Beehive, 200
belts (cloud bands), 127
Bennett, Jeffrey, 173
Beta Orionis, 193
Beta Persei, 191
Betelgeuse, 45, 193

Big Bang
  about, 251
  age of universe, 245, 258
  causes of "noise" in radio/TV receivers, 264
  competing theories, 268
  cosmic microwave background, 257
  distance between galaxies, 259–260
  evidence for, 252–253
  expansion of universe, 255–256, 259
  Hubble Flow, 29
  inflation, 253–255
  revision of theory, 176
  speed of galaxies, 258
  vacuum of space, 254
  what happened during, 246
Big Crunch, 255
Big Dipper, 48–49
big-light bending, 191
Billion-Channel Extraterrestrial Assay (Project
    BETA), 237
binary stars, 182–183, 293
binary systems, 123, 293
binoculars, 50–52, 165–166. *See also* eye safety
biosphere, 81
BL Lacs (BL Lacertae objects), 226
black holes
  about, 176, 217–218
  causes of, 175
  defined, 293
  diagramming, 181
  distortions of space/time caused by, 222–223
  inside parts of, 218–221
  outside parts of, 221
  quasars as, 224
  supermassive, 197, 226–227
  types of, 177, 218
blackouts, causes of, 151
blazars, 226
blue shift, 184
blue supergiants, 182
bolides, 62
Bootes Void, 214
boustrophedonic scanning, 73
brightness. *See also* luminosity; magnitudes
  determining, 181
  mathematics of, 22
  order of and naming conventions, 15, 16
  white dwarfs, 182
brightness scale. *See* magnitudes
British Astronomical Society, 32
bulge, 244
Burns, Joseph, 132
Butler, Paul, 240

• *C* •

Callisto, 129
Caloris basin, 100
canals on Mars, 103

Canis Major, 192
Canopus, 45
Capella, 45
Carina Nebula, 207
Cassegrain telescopes, 53
Cassini division, 132
Cassini space probe, 134
cataclysmic variables, 189
Catania Astrophysical Observatory, 163
celestial sphere, 26
Centaurus A, 212
Central Bureau for Astronomical Telegrams, 73
central peaks (Moon), 93
central stars of planetary nebulae, 173–174
centrifugal force, 148
Cepheid variables, 187–188, 259–260
Ceres, 122
CHANDRA Observatory, 221
Charon, 138, 142
chemical composition of stars, 185
Chicxulub crater, 119
chromosphere, 150
Ciel & Espace magazine, 34
circumpolar stars, 49
classes of stars, 21. *See also* magnitudes
classical novas, 189
classification of galaxies, 208
cloud bands (Jupiter's), 127
clouds, 127, 203. *See also* Large Magellanic
    Cloud; Small Magellanic Cloud
cluster galaxies. *See* galaxy clusters; star
    clusters
Coal Sack, 207
coeval, 182
cold dark matter, 245
color
  Neptune, 141
  red dwarf stars, 172
  of sky objects, 55
  temperature, diagramming, 178
  Uranus, 141
  on Venus, 101
color sphere, 150
coma, 70. *See also* comets
Coma Berenices, 244
Coma Cluster, 244
Comet IRAS-Iraki-Alcock, 71
comet tails, 263
cometary meteoroids, 60
comets
  announcements of upcoming, 27
  "comets of the century," 71
  defined, 293
  finding, 72–73
  versus meteors, 67
  naming conventions, 74
  occultations caused by, 123
  strikes on Jupiter, 131
  structure of, 67–70
  viewing, 70

comparative planetology, 83, 115
comparison charts, 194
comparison stars, 194
conjunctions, 106, 107–108
constellation designations, 45
constellations. *See also* by name, e.g. Ursa Major
    arrangement of Zodiac, 47
    early recognition of, 13
    list of, 17–20
    naming conventions, 13–14, 16–17
continental drift, 80, 102
convection zone, 148
Coordinated Universal Time (UTC), 84–85
Copernicus, Nicholas, 113
core, 82, 97, 149
corona, 150, 165
coronal mass ejections, 151–152
Cosmic Background Explorer satellite, 257
cosmic expansion, 29, 252
cosmic microwave background, 252, 253, 256, 257
Cosmic Perspective, The (Rennet, Donahue,
        Schneider, Voit), 173, 174
cosmic ripples, 257
cosmic voids, 214, 253
cosmologists, 245
cosmology, theories of, 176
Cowen, Ron, 249, 260
Crab Nebula, 174, 206
craters. *See* impact craters
critical density, 245, 255
cruises and tours, eclipse, 39–41
Crux, 13
cryosphere, 80
cryovolcanism, 137
curvature of universe, 254–255
cyberspace. *See* Web sites

● **D** ●

dark adaption, 65
dark matter, 243–248, 266, 293
dark nebulae, 203
daughter isotopes, 87
day. *See* rotation period (day)
daylight fireballs, 61
Daylight Saving Time, 85
death of a star, 205
declination circles, 25
declination (Dec), 24, 25, 106
deep sky objects, finding, 21
degenerate matter, 266
Deimos, 103
Deneb, 45
density, 88–89, 97, 100
detecting black holes, 176
diagramming the stars. *See* Hertzsprung-Russell
        (H-R) diagram
Dione, 134

Dirac, Paul, 249
distance measurements. *See also* measurements
    approximation of, 269
    Astronomical Units (A.U.s), 24
    galaxies, 259–260
    light-day, 224
    light-year, 22, 23, 269
    Milky Way from other galaxies, 198
    parsecs, 213
    standard candle, 259
    to stars, 13, 187, 259
distortions of space/time, 222–223, 247–248
Dobsonian mount (for telescopes), 55
Donahue, Megan, 173
Doppler, Christian, 184
Doppler Effect
    about, 184–186
    defined, 293
    using the, 191, 239, 294
Double Cluster, 200
double planets, 79, 138
double stars, 183, 293
Drake, Frank, 232–234
Drake's Equation, 232–234, 238
Dumbbell Nebula, 206
dust storms on Mars, 112
dust tail of comets, 68–69
dwarf galaxies, 210
dwarf novas, 189
dwarf stars. *See* red dwarf stars; white dwarf stars

● **E** ●

Earth. *See also* Moon (Earth's)
    age of, 87–88
    dark matter on, 266
    description of, 79
    escape velocity of, 218
    impact craters, 118
    movement of, 28–29, 47, 83–88
    properties of, 80
    regions of, 80–81
eclipse cruises and tours, 39–41
Eclipse Shades, 166. *See also* eye safety
eclipses. *See also* eye safety; occultations
    annular, 164
    binary stars, 187
    Jupiter's moons, 130
    Moon (Earth's), 90–92, 91
    partial, 92, 164, 165
    path of totality, 39, 91, 167–168
    stellar, 190–191
    Sun, 164, 165, 167–168
ecliptic, 47, 86, 293
ecliptic latitude, 106
Edgar Wilson Award, 73
Edmund Scientific, 94
Eight-Burst Nebula, NGC 3132, 207

Einstein, Albert
  gravity, theory of, 27, 28, 254
  Theory of General Relativity, 28, 191, 247
elliptical galaxies, 209
elongation, 106
Enchanted Skies Star Party, 39
end states of stellar evolution
  black holes, 176–177
  central stars of planetary nebulae, 173–174
  defined, 172
  neutron stars, 174–175
  supernovas, 175–176
  white dwarfs, 173
energy, 147–148, 154, 255–256
energy-generating core, 149
ephemeris, 27
equatorial mount (for telescopes), 55
eruptive prominences, solar, 152
escape velocity, 217–218
Eta Carinae, 190, 265–266
Europa, 129
evening stars, 105
event horizon, 218, 219–220
events
  annual meteor showers, 64
  best opposition of Mars, 111
  comet announcements, 27
  coming total eclipses of the Sun, 167–168
  eclipsing stars, 191
  halo event, 152
  locations of planets, 2,000-2,004, 273–283
  Mercury in transit, 114
  meteor showers, 62
  minima of Algol, 191
  occultations, 130
  tilting of Saturn's rings, 133
  total lunar eclipses, 91
  Venus in transit, 110
evolution. See end states of stellar evolution
expansion of universe, 176, 252
exploding stars, 187, 189–190
extra-solar planets
  continuing search for, 241
  51 Pegasi, 239–240
  how they're found, 238–239
  suitable for life, 240–241
  Upsilon Andromedae system, 240
extraterrestrial life. See SETI (Search for
    Extraterrestrial Intelligence)
extrinsic variable stars, 187
eye safety
  importance of, 145, 146
  lunar eclipse viewing, 91
  projection technique, 156–158
  solar eclipse viewing, 165–166
  solar filters, 56, 159–161, 166
  white light flare viewing, 188–189

• F •

"face" on mars, 104
face-on spiral galaxies, 211
falling stars. See meteors
false color, 55
far side of Moon (Earth's), 94
festoons, 127
51 Pegasi, 239–240
film speed, 66
filters, solar. See eye safety
filters for white light flare viewing, 188–189
finding galaxies, 21
finding planets, 241, 273–283
fireballs, 61, 62
first contact (solar eclipse), 165
"fixed stars," 24
flare stars, 187, 188–189
following spot (sunspots), 153
Fomalhout, 45
Ford, Kent, 244
formation of Milky Way, 199
47 Tucanae, 202
fossil evidence on Mars, 104
Foucault pendulums, 46
fourth contact (solar eclipse), 167
frequency of sound/light, 184
front-end filters, 159
full Moon, 90, 95
full-aperture filters, 159
funny energy, 255–256
fusion, nuclear, 146, 171

• G •

galactic bulge, 197
galactic center, 197
galactic coordinates, 198
galactic disk, 196
galactic equator, 198
Galactic Latitude, 198
Galactic Longitude, 198
galactic nuclei, 224–227
galactic plane, 198, 214
galactic rim, 198
galactic year, 28, 198
galaxies. See also by name, e.g. Pinwheel Galaxy
  centers of, 219
  classification of, 208
  defined, 293
  distance between, 259–260
  dwarf, 210
  face-on spiral, 211
  finding, 21
  formation of, 257
  guides for finding, 21
  host galaxy, 225
  images of on Web, 214–215

interarm regions of, 196
irregular/dwarf/low surface, 210
light from, 255
list of best, 211–212
Local Group of Galaxies, 29, 210, 212–213
low surface brightness, 210
radio, 226
seeds of galaxy formation, 257
shape, 196, 208
in Southern Hemisphere, 212
speed of, 244, 258
spiral, 196, 208, 244
superclusters/great walls/cosmic voids, 214
types of, 207–210
galaxy clusters
    causes of, 253
    defined, 210
    ghost images of, 248
    Local Supercluster, 214
    rich clusters, 213
    superclusters, 214
    Virgo Cluster, 195, 213
    Virgo Supercluster, 214
Galilean moons, 129–131
Galileo, 132, 146
gamma rays, 249
Gamow, George, 252
Ganymede, 129
gas, 13, 150
gas-giant planets, 126
Geminids meteor showers, 63, 64
general relativity theory. *See* Theory of General
    Relativity
geology of Moon, 92–95
geomagnetic field, 82, 83
geomagnetic storms, 151
German Space Operations Centre (GSOC), 76
ghost images of clusters, 248
Giant Impact theory, 96
giant molecular clouds nebulae, 203
gibbous Moon, 90
global magnetic fields, 82, 100
Global Positioning System (GPS), 39, 123
globular clusters, 188, 201–202
GONG project, 163
"good" horizon, 46
good seeing, 51, 112
GPS (Global Positioning System), 39, 123
gravitational lensing, 191, 248
gravity
    of asteroids, 118
    of black holes, 176, 221
    Jupiter/Saturn, 126
    of Milky Way, 199
    repulsive, 254
    Sun's, 148
    surface, 185
    theories of, 13, 27–28, 254
Great Comet of 1910, 71
Great Dog. *See* Sirius

Great Red Spot, 125, 127–128
Great Spiral Galaxy, 211
Great Walls, 195, 214
great white storm (Saturn's), 133
greatest elongations, 106
Greek alphabet, 15
greenhouse effect on Venus, 115
Greenwich Mean Time, 84
Griffith Observatory and Planetarium, 36, 37
GSOC (German Space Operations Centre), 76
Guth, Alan, 253

Hadar, 45
Hadley Rille, 93
Hale-Bopp Comet, 67, 71
half Moon, 90
half-life, 87
Halley's Comet, 67, 71
halo event, 152
halo of dark matter, 245
H-alpha filters, 159
Hayden Planetarium, 38
heliocentric (Sun-centered) theory, 113
helium, 252–253
Heraclides, 46
Herbig-Haro objects, 171
Hertzsprung-Russell (H-R) diagram, 178–182
HII regions, 171, 203
Hipparchos (or Hipparchus), 21
horizon, "good," 46
host galaxy, 225
hot Jupiters, 239, 240
H-R diagram, 178–182
Hubble, Edwin P., 204, 208, 252
Hubble age, 258
Hubble Constant, 258–260
Hubble Flow, 29
Hubble Telescope, 22, 73, 270
Hubble types, 208
Hubble's Law, 213
Hyades, 200
Hyakutake Comet, 71
Hydra, 13
hydrogen atoms, 206–207
hydrogen gas, 235
hydrogen-burning shell, 172
hydrosphere, 80
Hygiea, 122

• I •

IAU (International Astronomical Union), 13, 73,
    119, 134
IC (Index Catalogue), 21
icy planets, 135–136

Ida, 117
Ikeya-Seki, Comet, 71
impact basins, 100, 129
impact craters
  defined, 293
  Earth's, 119
  impact basins, 100
  Mars, 103
  Moon (Earth's), 93, 96, 118
  Venus, 101
in transit. *See* transits
Index Catalogue (IC), 21
inferior conjunctions, 107–108
inflation of universe, 253–255
interarm regions of galaxies, 196
intermediate-mass black holes, 177
intermediate-mass stars, 172
International Astronomical Union (IAU), 13, 73, 119, 134
International Meteor Organization, 67
International Space Station, 75
intrinsic variable stars, 186–187
inverse square law, 187
Invisible Universe, The (Malin), 199
Io, 129
ion tail, 70. *See also* comets
ionization state, 185
IOTA (International Occultation Timing Association), 123
IRAS-Iraki-Alcock comet, 71
Iridium communication satellites, 75, 76
iron meteorites, 60
irregular galaxies, 210
isotopes, 87

• J •

jets from radio sources, 224
Jewel Box, 200
Jim Hendrick Studio, 159
Jupiter
  about, 125–126
  escape velocity of, 218
  finding, 126
  future locations of, 273–283
  Great Red Spot, 125, 127–128
  moons, 129–131
  rings, 127–128
Jupiter probe, 131

• K •

KBOs (Kuiper Belt Objects), 139–140
Keck Observatory, 41, 118
Kepler's Star, 190
killer asteroids, 267–268. *See also* Near Earth Objects (NEOs)
Kirschner, Robert, 214

Kitt Peak National Observatory, 36, 37
Kronk, Gary, 67
Kuiper Belt Objects (KBOs), 139–140

• L •

Lagoon Nebula, 207
Large Magellanic Cloud, 192, 197, 212, 248
laser altimeters, 102
last contact (solar eclipse), 167
last quarter Moon, 90
latitude, ecliptic, 106
lava beds, 93
Lawrence Livermore National Laboratory, 134
layers of Sun, 149
leading spot (sunspots), 153
Learmonth Solar Observatory, 163
Leavitt, Henrietta, 187, 260
lensing, 191
lenticular galaxies, 209
Leonids meteor showers, 64
Levy, David, 72, 131
life cycles of stars
  about, 169–170
  death of a star, 205
  end states of stellar evolution, 172–177
  main sequence stars, 171–172
  normal stars, 170
  red giants, 172
  young stellar objects (YSOs), 171
life on other planets. *See* SETI (Search for Extraterrestrial Intelligence)
lifespan of Sun, 155–156
light. *See also* magnitudes
  annihilation radiation, 249
  from black holes, 176, 218
  from galaxies, 255
  from Large Magellanic Clouds, 248
  recognizing celestial objects, 11–12
  from universe, 257
  wavelength, 184
light-day, 224
light-years, 22, 23, 269
limiting magnitude, 48
Lincoln Near Earth Asteroid Research (LINEAR) project, 121
liquid water, 80, 96, 102–103
lithosphere, 80
Little Dipper, 48
Local Group of Galaxies, 29, 210, 212–213
Local Supercluster, 214
Loch Ness Productions, 38
long period variable stars, 188
longitude, 106
low surface brightness galaxies, 210
Lowell, Percival, 36, 37, 103, 264
Lowell Observatory, 36, 37
Lowell's Theory, 264
lucida, 16

luminosity, 147, 154, 180, 188, 259
lunar charts, 94
lunar cycle, 90
lunar eclipses, 90–92
lunar highlands, 93
lunar mountains, 94
lunar occultations, 123
Lunar Prospector, 97
lunation, 90
Lyra, 193, 204–205

• *M* •

MACHOs (massive compact halo objects), 246, 247
magazines
  Astronomy, 33, 34, 130
  Ciel & Espace, 34
  Sky & Telescope, 34, 130
  SkyNews, 34
  Web sites, 33–34
Magellanic Clouds, 192, 197, 212, 248
magnetic cycle, Sun's, 154
magnetic fields
  Earth's, 82
  Jupiter's/Saturn's, 126
  Mars, 102–103
  Moon (Earth's), 97
  revealed by spectroscopy, 185
  Sun's, 148
magnetic reconnection, 151
magnetized rock, 83
magnetographs, 151
magnetosphere, 81, 150
magnification of telescope eyepieces, 54
magnitude, 21–22, 48, 178, 188
magnitudes
  brightest stars seen from Earth, 45
  Pluto, 142
  of stars in constellations, 17–20
main sequence stars, 171–172, 181, 182
Maksutov-Cassegrains telescopes, 53, 57, 156
Malin, David, 199
Malin Space Systems, 102
mantle, 100
mapping dark matter, 248
Marcotte, Claude, 33
Marcy, Geoff, 240
maria, Moon's, 93
Mariner 4 spacecraft, 104
Mariner 10 spacecraft, 99, 100
Mariner Valley (Valles Marineris), 103
Mars
  atmosphere, 115
  best times to view, 110–113
  escape velocity of, 218
  future locations of, 273–283
  how to find, 104–105
  life on, 103–104
  moons, 103
  motion of, 110–113
  opposition of, 106
Mars Global Surveyor (MGS), 102
Mars Pathfinder, 102
Mars Watch site, 112
mass
  classifying by, 180
  distortions of fabric of space by, 247–248
  intermediate mass black holes, 177
  intermediate-mass stars, 172
  life cycles and, 170
  of planets, 126
  relative, 184, 185
  of universe, 255
  visible, 245
massive compact halo objects (MACHOs), 246, 247
matter, 243–247, 249
Mauna Kea Observatories, 36, 37
maxima, 191
Mayor, Michael, 239
Meade ETX-90/EC telescope, 57, 156
mean solar time, 84
measurements. *See also* distance measurements
  apparent size, 109
  of black holes, 220
  light years, 269
  minute of arc, 25
  minutes of time, 25
  Right Ascension/Declination, 24
Mega-Channel Extraterrestrial Array (Project META), 237
Mercury, 99–100, 104–105, 113, 114
MESSENGER (MErcury Surface, Space ENvironment, GEochemistry and Ranging), 99, 100
Messier 13, 202
Messier 15, 202
Messier 32, 211
Messier 33, 211
Messier 49, 213
Messier 51, 211
Messier 87, 213
Messier 94, 211
Messier 100, 213
Messier, Charles, 21
Messier Catalog, 21
Messier Catalog Web Site of Students for the Exploration and Development of Space, 21
Messier Objects, The (O'Meara), 21
Meteor Crater, 118
meteor showers
  about, 61, 62
  from comet dust, 70
  naming conventions, 63
  occasions of, 62
  photographing, 66
meteor train, 62

meteorites
  defined, 59, 294
  from Mars, 104, 263–264
  naming conventions, 74
  ownership of, 74
  radioactive dates, 88
  temperature of, 268
  types of, 60
meteoroids, 59, 60, 118, 294
meteors
  versus comets, 67
  correct terminology, 59–60
  defined, 294
  how to observe, 65–66
  naming conventions, 74
  photographing, 66
  tracking, 64
MGS (Mars Global Surveyor), 102
microlensing events, 187, 191–192
micrometeorites, 264
Microwave Anisotropy Probe (MAP) satellite, 256
microwave background radiation, 235, 252, 264
Milky Way, 45, 195–199
minima, 191
Minima of Algol, 191
Minor Planet Center, 119, 121
minor planets. *See* asteroids
' minute of arc, 25, 109
minutes of time, 25
Mira stars (or variables), 188
MIRV, 120
molecular clouds, 203
Molokai Ranch, 41–42
Moon (Earth's), 97. *See also* lunar entries
  density, 88–89, 97
  eclipses, 90–92
  geology, 92–95
  impact craters, 118
  locating, 273
  lunar charts, 94
  lunar cycle, 90
  Lunar Prospector, 97
  maps, 94
  motion of, 94
  occultations caused by, 123
  ocean tides on Earth, 265
  origin of, 95–97
  phases, 89–90
moon shadows, 130
moons
  of asteroids, 117
  Earth's, 88–97
  Jupiter's, 129–131
  Mars's, 103
  Neptune's, 137, 142
  planets with no, 100, 101
  Pluto's, 138, 142
  Saturn's, 133–134
  types of, 134
  Uranus's, 136, 141

Morning Star, 269
morning stars, 104–105
motion of sky objects
  artificial satellites, 75
  asteroids, 118
  Earth, 28–29, 83–88
  Mars, 110–113
  Moon (Earth's), 94
  retrograde, 111
  Uranus, 141
Mount Wilson Observatory, 36, 37, 163
mounts, telescope, 55
Mt. Evans Meyer-Womble Observatory, 36, 37
multiple stars, 182–186, 184
multiverse, 253
Mysteries of Deep Space (television series), 38
mythology, science versus, 13

• *N* •

naked-eye observation, 44–46
naming conventions, 13–14, 16–17, 63, 74
NASA Web sites
  "Ask a Scientist," 33
  current solar images, 163
  eclipses, 168
  main, 102
  Microwave Anisotropy Probe (MAP)
    satellite, 256
  SOHO Movie Theater, 164
  Web sites, 256
National Museum of Science and Technology
    Corporation, 34
National Optical Astronomy Observatories, 36, 37
National Radio Astronomy Observatory, 36–37,
    37–38
National Solar Observatory, 36, 37
NCP (North Celestial Pole), 25
Near-Earth Asteroid Tracking (NEAT) project,
    121
Near Earth Objects (NEOs), 119–120, 121,
    267–268. *See also* Potentially Hazardous
    Asteroids (PHAs)
NEAT (Near-Earth Asteroid Tracking) project,
    121
NEB (North Equatorial Belt), 127
Nebraska Star Party, 39
nebulae. *See also* by name, e.g. Crab Nebula;
    supernovas
  central stars of planetary, 173–174
  defined, 294
  early thinking about, 204
  guides for finding, 21
  images of on Web, 215
  list of important, 206–207
  planetary, 53, 155–156, 204–205, 294
  in Southern Hemisphere, 207
  types, 203–204
negative curvature, 254–255

negative magnitude, 22
NEOs (Near Earth Objects), 119–120, 121, 267–268
Neptune, 135–137, 141–142
neutrino observatory (AMANDA), 154
neutrinos, 154–155, 246
neutron stars, 174–175, 181, 294
New General Catalogue (NGC), 21
New Moon, 90
Newton, Isaac, 13, 27, 28
Newtonian reflectors, 53, 156
NGC 205, 211
NGC 253, 212
NGC 5128, 212
NGC 6231, 200
NGC 7000, 206
NGC (New General Catalogue), 21
normal stars, 172
North American Meteor Network, 67
North American Nebula, NGC 7000, 206
North Celestial Pole (NCP), 25
North Equatorial Belt (NEB), 127
north-seeking compass, 153
North Star (Polaris), 47–49
Northern Coal Sack Nebula, 206
Northern Lights (aurora borealis), 82, 98, 151
Norton, Arthur P., 58, 112
Norton, Peter, 202
Norton's Star Atlas and Reference Handbook (Epoch 2000.0) (Norton), 58, 112, 202
Nova Search program, 194
novas, 189. *See also* nebulae; supernovas
nuclear fusion, 146, 171
nucleus of comets, 67
numbers of moons, 134

• *O* •

OB associations, 202
objects, sky, 12, 13, 21, 220. *See also* motion of sky objects
oblate shape, 128
observation, 43–46, 65, 193–194. *See also* viewing, best times for; viewing, worst times for
observational data, 180–181
observatories
  AMANDA (neutrino observatory), 154
  Catania Astrophysical Observatory, 163
  CHANDRA Observatory, 221
  Griffith Observatory and Planetarium, 36, 37
  Keck Observatory, 41, 118
  Kitt Peak National Observatory, 36, 37
  Learmonth Solar Observatory, 163
  Lowell Observatory, 36, 37
  Mauna Kea Observatories, 36, 37
  Mount Wilson Observatory, 36, 37, 163
  Mt. Evans Meyer-Womble Observatory, 36, 37
  National Optical Astronomy Observatories, 36, 37
  National Radio Astronomy Observatory, 36–37, 37–38
  National Solar Observatory, 36, 37
  orbiting, 221
  Palomar Observatory, 36, 37
  Solar and Heliospheric Observatory (SOHO), 164
  U.S. Naval Observatory, 36, 37
  visiting, 36–37
  Web sites, 37
Observer's Handbook of the Royal Astronomical Society of Canada, 34
occultations, 123, 130, 294. *See also* eclipses
ocean tides, 265
off-axis filters, 159
Olympus Mons, 103
O'Meara, Stephen J., 21
Omega Centauri, 202
Omicron Ceti, 188
open clusters, 199–200, 202
oppositions, 106, 111
optical double stars, 183
optically violently variable quasars (OVVs), 225
orbiting observatories, 221
orbits
  binary stars, 183
  determining speed of, 191
  Earth's, 28, 84–85
  Hubble Telescope, 270
  Jupiter's moons, 130
  Mercury, 106
  Moon (Earth's), 89–90
  Neptune's moons, 137
  outlying stars in galaxies, 244
  Pluto, 137, 142
  Saturn, 133
  solar system around galactic center, 198
  types of, 134
  Uranus's moons, 136
ordinary novas, 189
origins of Moon, 95–97
Orion, 49
Orion Molecular Cloud, 203
Orion Nebula, 171, 203, 206
Orion Telescopes & Binoculars, 94
orti, 294
OVVs (optically violently variable quasars), 225
oxygen atmosphere, 80

• *P* •

Pallas, 122
Palomar Observatory, 36, 37
parsecs, 213
partial eclipses, 92, 164, 165
particles, 246–247, 249
path of totality, 39, 91, 167–168. *See also* eclipses

patterns of stars. *See* asterisms, constellations
penumbra, 161, 165
Penzias, Arno, 252
Period-Luminosity relation, 187
Perseids meteor shower, 63
PHAs (Potentially Hazardous Asteroids), 119–120, 121
phases of Moon, 89–90
phases of Venus, 108–110
Phobos, 103
photinos, 246, 247
photography, 54, 66, 103
photons, 247
photosphere, 149
Pinwheel Galaxy, 211
planet happenings, 2000-2004, 273–283
planetarium software, 34–35
planetariums, 34–35, 36, 37, 38. *See also* observatories
planetary nebulae, 53, 155–156, 204–205, 294
planetary occultations, 123
planetary rings, 123, 131–134, 141
planets, 12–13, 56, 273–283, 294. *See also* by name, e.g., Jupiter
planispheres, 47
plasma, 150
plasma tail, 70. *See also* comets
plate tectonics, 80, 102
Pleiades, 200
Plutinos, 139–140
Pluto, 137–139, 142, 264
polar alignment, 55
Polaris, 47–49
polarity of sunspots, 153
polarization, 226
Pollux, 45
Polymer Plus filters, 161
positive curvature, 255
positrons, 249
Potentially Hazardous Asteroids (PHAs), 119–120, 121. *See also* Near Earth Objects (NEOs)
Procyon, 45
prograde orbit, 134
Project BETA (Billion-Channel Extraterrestrial Assay), 237
Project META (Mega-Channel Extraterrestrial Array), 237
Project Ozma, 232–233
Project Phoenix, 234–236
projection technique, 156–158, 188–189. *See also* eye safety
prominences, solar, 152
proportionality, 258
protection from Sun. *See* eye safety
protons, 249
protoplanetary nebulae, 205
Proxima Centauri, 172, 189, 192
publications. *See* magazines
pulsars, 174–175, 294
pulsating stars, 187–188

• Q •

QSOs (quasistellar objects), 225
Quadrantids meteor showers, 63–64
quadruple stars, 183, 206
quarks, 247
quarter Moon, 90
Quasar PKS2349, 223
quasars, 223–224, 294
Queloz, Didier, 239

• R •

radar maps, 101
radial velocities, 185
radiant, 62
radiation, 249, 252. *See also* cosmic microwave background
radiation belts, Earth's, 82
radiative zone, 149
radio galaxies, 226
radio-loud quasars, 225
radio-quiet quasars, 223, 225
radio sources, 224
radio telescopes, 177, 198, 232–233, 234
radio waves, 234, 264
radioactive dating, 87
radioactive isotopes, 87, 96
rain on Venus, 101
rays, 93
Realm of the Nebulae, The (Hubble), 204
recording observations, 46, 61–62
red dwarf stars
    color of, 172
    determining brightness, 182
    flare stars, 188–189
    hydrogen burning rate, 170
    luminosity class of, 180
red giant stars, 172, 294
red supergiants, 172, 182
redshift, 186, 294
reflecting telescopes, 53, 54
reflection nebulae, 203
refracting telescopes, 53, 156
regions of Earth, 80–81
regions of Sun, 148–150
regions of universe, 253
regular moons, 134
relative mass, 185
relativity theory. *See* Theory of General Relativity
repulsive gravity, 254
retrograde motion, 111
retrograde orbits, 137, 141
revolution period (year), Uranus's, 136
revolutions, Pluto's moon, 142
revolutions around the Sun, Earth's, 28
Rhea, 134

rich clusters, 213
Rigel, 45, 193
right ascension (RA), 24, 25, 106
Rigil Kentaurus, viewing, best times for, 45
rilles, 93, 101
Ring Nebula, 204–205, 206
Ring Nebula in Lyra, 173
rings, planetary, 123, 131–134, 141
Riverside Amateur Telescope Makers (star
    party), 39
Roger W. Tuthill, Inc., 160
rotation, 94, 294
rotation on axis, Earth's, 28
rotation period (day)
  Earth's Moon, 29
  Jupiter's, 126–127
  Pluto's, 138
  proof of Earth's, 46
  Uranus, 136
rotational axis, 139
Royal Astronomical Society of Canada, Web site, 32
RR Lyrae stars, 188
Rubin, Vera, 244

● **S** ●

safety. *See* eye safety
Sagan, Carl, 11, 233
Sagittarius A*, 177, 197
Sagittarius Dwarf Galaxy, 212
satellites, artificial, 73–75, 102
Saturn, 125–126, 131–134, 273–283
Schmidt-Cassegrains telescopes, 53, 57, 156
Schneider, Nicholas, 173
science, versus mythology, 13
SCP (South Celestial Pole), 25
Sculpter Galaxy, 212
search for dark matter, 247–248
Search for Extraterrestrial Intelligence (SETI), 11
Search for Extraterrestrial Radio Emissions from
    Nearby Developed Intelligent Populations
    (SERENDIP), 237, 238
searching for extraterrestrial life. *See* SETI
    (Search for Extraterrestrial Intelligence)
seasons, causes of Earth's, 85–87, 268–269
SEB (South Equatorial Belt), 127
second contact (solar eclipse), 165
second of time, 25
" second of arc, 25, 27, 109
seeds of galaxy formation, 257
seeing, 112, 294. *See also* atmospheric turbulence
seeing (atmospheric conditions), 51
SERENDIP (Search for Extraterrestrial Radio
    Emissions from Nearby Developed
    Intelligent Populations), 237, 238
SETI League, 237
SETI (Search for Extraterrestrial Intelligence)
  about, 11
  current searches, 234–238

Drake's Equation, 232–234
likelihood of other life, 231–232, 233, 240
massive stars, 173
other projects, 237–238
searching yourself, 238
Web sites, 237
Seyfert, Carl, 225
Seyfert galaxies, 225
shadow bands, 166
shape
  of asteroids, 268
  elliptical, 209
  galaxies, 196, 208
  Milky Way, 196–198, 199
  oblate, 128
  Sun's, 147–148
  of universe, 254–255
Shoemaker-Levy 9, 131
shooting stars. *See* meteors
Shostak, Seth, 241
sidereal day, 84
signals, searching for extraterrestrial, 235–236
singularity, 218, 220
Sirius, 22, 45, 192–193
Sirius B, 193
size
  apparent, 109
  of asteroids, 118
  Milky Way, 199
  Saturn's rings, 132
  of stars, determining, 191
  Sun, 147–148
Sky & Telescope magazine, 33, 34, 130
sky darkness, 112
SkyNews magazine, 34
Skywatcher's Inn, 41
Small Magellanic Cloud, 197, 212
software, recommended, 34–35
SOHO Movie Theatre Web site, 164
SOHO (Solar and Heliospheric Observatory), 164
Sojourner, 102
solar activity, 151–154
Solar and Heliospheric Observatory (SOHO), 164
solar constant, 154
solar cycles, 151–154
solar disk, 157
solar eclipses, 90, 150, 164–168
solar filters, 56, 159–161. *See also* eye safety
solar flares, 151–152
solar luminosity, 154
solar magnetism, 151
solar neutrino deficiency, 155
solar prominences, 159
solar rotation, 157
Solar Skreen Sun Filters, 160
solar wind, 70, 150
Sombrero Galaxy, 211
sound wavelength, 184
South Celestial Pole (SCP), 25
South Equatorial Belt (SEB), 127

Southern Cross, 13, 200
Southern Hemisphere, viewing in
  galaxies, 212
  globular clusters, 202
  nebulae, 207
  Southern Cross, 13, 200
  Southern Lights (aurora australis), 82, 151
  star clusters, 200
Southern Lights (aurora australis), 82, 151
Southern SERENDIP, 237
space weather, 151
Spaceguard Foundation, 121
spectral class, 45, 179
spectral lines, 185, 186, 224
spectral types, 178–179
spectrographs, 13
spectroscopy, 185
spectrum, 178
Spica, 45
spiral galaxies, 196, 208, 244
sporadic meteors, 61
squarks, 246, 247
standard candle, 259
Standard Time, 85
star clusters
  defined, 295
  distance of, 188
  Double Cluster, 200
  globular, 201–202
  guides for finding, 21
  versus multiple star system, 184
  OB associations, 202
  open clusters, 199–200, 202
  in Southern Hemisphere, 200
Star Hill Inn, 41
star names, purchasing, 256
star parties, 38–39
star party (Riverside Amateur Telescope
  Makers), 39
Stark Effect, 185
Starry Night Deluxe (software), 35
stars
  versus asteroids, 122
  brightest as seen from Earth, 45
  brightest in constellations, 17–20
  defined, 295
  distances of, 13
  naming conventions, 13–14, 16–17
  versus planets, 12–13
Stellafane (star party), 39
stellar eclipses, 190–191
stellar evolution. *See* end states of stellar
  evolution
stellar interior, 148
stellar mass black holes, 177, 218, 219
stellar nurseries, 171
stellar spectroscopy. *See* spectroscopy
stellar wobble, 239
stony-iron meteorites, 60
stony meteorites, 60

stopping down, 160
storms, 112, 133, 151
straight line distortions, 222
Students for the Exploration and Development
  of Space, 21
subatomic particles, 247, 249
subgiants, 180
summer solstice, 87
Summer Time, 85
Sun. *See also* solar entries
  about, 146–147
  compared to other stars, 270
  core, 149
  energy, 147–148, 154
  eye safety for viewing, 145, 146, 156–161,
    165–166
  gravity, 148
  heliocentric theory, 113
  layers of, 149
  lifespan of, 155–156
  luminosity class of, 180
  magnetic cycle, 154
  magnetic fields, 148
  neutrinos, 154–155
  regions, 148–150
  size/shape, 147–148
  solar eclipses, 90, 150, 164–168
  sunspots, 149, 152–154, 161–163
Sun-centered (heliocentric) theory, 113
sunlike stars, 172
sunspot cycles, 152–154, 185
sunspot numbers, 162–163
sunspots, 149, 161–163, 265
superclusters, 214
supergiants, 172, 180, 182
superior conjunction, identifying, 107–108
superluminal motion, 224
supermassive black holes, 177, 197, 218, 219,
  226–227
Supernova 1987A, 197
Supernova Search program, 194
supernovas. *See also* nebulae
  about, 189–190
  brightness of, 255
  defined, 295
  nebulae of, 189
  remnants of, 204, 205
  Type Ia, 176, 189, 255, 256, 260
  Type II, 175
  views of, 265–266
surface gravities, 185
synchronous rotation, Moon (Earth's), 94
Syrtis Major, 111

• *T* •

T Tauri stars, 171
Tarantula nebula, 197, 207
telescope motels, 41–42

telescopes. *See also* eye safety; radio telescopes
   classifications of, 53
   for comet viewing, 72
   economical ways to buy, 56–57
   Hubble Telescope, 73, 270
   projection techniques, 156–158
   radio telescopes, 177, 198, 232–233, 234
   types of mountings, 55
   upside down view in, 55
   Web sites, 22, 41–42, 57, 121
   X-ray, 221
temperature
   cosmic microwave background, 252
   diagramming, 178
   Jupiter/Saturn, 126
   Mercury, 100
   meteorites, fallen, 268
   Moon (Earth's), 96
   Pluto, 138
   surface of types of stars, 179
terminator, 94–95, 109, 295
terrestrial planets, 79, 104–115
Texas Star Party, 38, 39
theories. *See also* Big Bang
   accelerating universe, 255
   Giant Impact, 96
   of gravity, 13, 27–28, 254
   heliocentric (Sun-centered), 113
   inflation of universe, 254–255
   Lowell's Theory, 264
   Theory of General Relativity, 28
   Unified Model of Active Galactic Nuclei, 227
Theory of General Relativity, 28, 191, 247, 258
TheSky (software), 35
third contact (solar eclipse), 167
Thousand Oaks Optical, 160–161
tides, Earth's ocean, 265
tilted planets, 136–137, 139
time, 25, 84–85
Titan, 133–134
Tombaugh, Clyde, 139
total eclipses, 91, 164, 165. *See also* eclipses; path of totality
tours and cruises, eclipse, 39–41
transition region, 150
transits
   defined, 56, 295
   Jupiter's moons, 130
   Mercury, 114
   Venus, 110
transparency (atmospheric), 112
Trapezium, 206
Triangulum Galaxy, 197, 211
Trifid Nebula, 207
triple stars, 183, 192
Triton, 137, 142
turbulence, atmospheric, 51
Turner, Michael, 255
Type Ia supernovas, 176, 189, 255, 256, 260
Type II supernovas, 175

• **U** •

UFOlogists, 11
ultraviolet light, 109–110
umbra, 91, 161, 165
Unified Model of Active Galactic Nuclei theory, 227
Universal Time (UT), 84
universe
   age of, 245, 258
   birth of, 253
   expansion of, 176, 252
   light from, 257
   shape of, 254
   uniformity of, 245
unpolarized waves, 226
Upsilon Andromedae system, 240
Uranus, 135–137, 140, 141
Ursa Major, 49
U.S. Census Bureau, 76
U.S. Gazetteer, 76
U.S. Naval Observatory, 36, 37
UT (Universal Time), 84
UTC (Coordinated Universal Time), 84–85

• **V** •

vacuum, 254
Valhalla, 129
Valles Marineris (Mariner Valley), 103
Van Allen, James, 82, 150
Van Allen Belts, 82, 150
variable stars, 186–187, 295
Vega, 45, 193
velocity, 184, 217–218, 258, 259
Venus
   about, 101–102
   best times to view, 107
   future locations of, 273–283
   greenhouse effect on, 115
   how to find, 104–105, 108–110
   magnitude of, 22
   rain on, 265
   in transit, 110
vernal equinox, 87
Very Long Baseline Array, 177, 198
Vesta, 118, 122
viewing, best times for. *See also* observation
   Andromeda Galaxy, 211
   based on planets' positions, 106
   Jupiter's moons, 130
   Mercury/Venus/Mars, 104–105
   Moon (Earth's), 95
   Uranus, 140–141
viewing, worst times for, 107
viewing the Sun. *See* eye safety
Viking Landers, 104
Viking Orbiters, 104

virga, 101
Virgo Cluster, 195, 213
Virgo Supercluster, 214
visible mass, 245
visible matter, 245
vision, dark adapting, 65. *See also* eye safety
Voit, Mark, 173
volcanism, 101–102, 103
volcanoes, 80
Voyager, 131

• W •

waning crescent Moon, 90
water, liquid, 80, 96, 102–103
Water Snake, 13
wavelength of sound/light, 184
waxing crescent moon, 90
weakly interacting massive particles (WIMPs), 247
weather, 101
Web sites
  accelerating universe theory, 255
  ALPO, 114
  American Association of Variable Star
    Observers (AAVSO), 194
  asteroid discovery telescopes, 121
  astronomy clubs, 32–33
  astronomy magazines, 33
  Cassini space probe, 134
  Central Bureau for Astronomical Telegrams, 73
  comets, 71
  Edgar Wilson Award, 73
  galactic images, 214–215
  halo events, 152
  help for novices, 194
  Hubble telescope photos, 22
  International Occultation Timing Association
    (IOTA), 123
  Jupiter's moons, 131
  Keck Observatory, 118
  Kuiper Belt Objects, 140
  lunar eclipses, 91
  lunar maps/charts, 94
  MAP satellite, 256
  Mars images, 102
  Mars Watch, 112
  MarsNet, 112
  MESSENGER, 100
  Messier Catalog Web Site of Students for the
    Exploration and Development of Space, 21
  meteor showers, 67
  Milky Way's galactic plane, 214
  NASA eclipse site, 168
  Neptune/Neptune's moons, 142
  observatories, 37
  Observer's Handbook of the Royal
    Astronomical Society of Canada, 34
  planetarium software, 35

Pluto, 139
Pluto finder chart, 142
position of Uranus/Neptune, 142
Potentially Hazardous Asteroids (PHAs), 121
satellite viewing predictions, 75–76
Saturn's moons, 134
search for extra-solar planets, 241
SETI programs, 238
solar filter sources, 159, 160, 166
solar pictures, 163–164
space weather reports, 151
star clusters, 199
star parties, 39
sunspot numbers, 162
telescope motels, 41–42
telescope product information, 57
travel agents, 41
Uranus images, 141
UTC time, 85
Venus images, 101
viewing guides, 113
Views of the Solar System, 118
weight, 173
West, Comet, 71
Whirlpool Galaxy, 211
white dwarf stars
  brightness, 182
  classification of, 180
  defined, 156, 295
  end states of stellar evolution, 173
white holes, 221
white light, 163
white light flares, 188–189
Wilson, Robert, 252
WIMPs (weakly interacting massive particles), 247
winter solstice, 87
wormholes, 220–221

• X •

X rays, 221, 224, 249

• Y •

YSOs (young stellar objects), 170, 171, 181

• Z •

Zeeman Effect, 185
zenith, 295
0 magnitude, 22
Zodiac, 47
zones (cloud bands), 127
Zwicky, Fritz, 244

FOR DUMMIES®

The easy way to get more done and have more fun

PERSONAL FINANCE & BUSINESS

0-7645-2431-3

0-7645-5331-3

0-7645-5307-0

Also available:

Accounting For Dummies
(0-7645-5314-3)

Business Plans Kit For Dummies
(0-7645-5365-8)

Managing For Dummies
(1-5688-4858-7)

Mutual Funds For Dummies
(0-7645-5329-1)

QuickBooks All-in-One Desk Reference For Dummies
(0-7645-1963-8)

Resumes For Dummies
(0-7645-5471-9)

Small Business Kit For Dummies
(0-7645-5093-4)

Starting an eBay Business For Dummies
(0-7645-1547-0)

Taxes For Dummies 2003
(0-7645-5475-1)

HOME, GARDEN, FOOD & WINE

0-7645-5295-3

0-7645-5130-2

0-7645-5250-3

Also available:

Bartending For Dummies
(0-7645-5051-9)

Christmas Cooking For Dummies
(0-7645-5407-7)

Cookies For Dummies
(0-7645-5390-9)

Diabetes Cookbook For Dummies
(0-7645-5230-9)

Grilling For Dummies
(0-7645-5076-4)

Home Maintenance For Dummies
(0-7645-5215-5)

Slow Cookers For Dummies
(0-7645-5240-6)

Wine For Dummies
(0-7645-5114-0)

FITNESS, SPORTS, HOBBIES & PETS

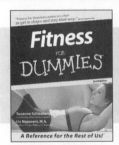

0-7645-5167-1

0-7645-5146-9

0-7645-5106-X

Also available:

Cats For Dummies
(0-7645-5275-9)

Chess For Dummies
(0-7645-5003-9)

Dog Training For Dummies
(0-7645-5286-4)

Labrador Retrievers For Dummies
(0-7645-5281-3)

Martial Arts For Dummies
(0-7645-5358-5)

Piano For Dummies
(0-7645-5105-1)

Pilates For Dummies
(0-7645-5397-6)

Power Yoga For Dummies
(0-7645-5342-9)

Puppies For Dummies
(0-7645-5255-4)

Quilting For Dummies
(0-7645-5118-3)

Rock Guitar For Dummies
(0-7645-5356-9)

Weight Training For Dummies
(0-7645-5168-X)

Available wherever books are sold.
Go to www.dummies.com or call 1-877-762-2974 to order direct

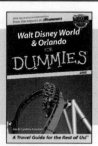

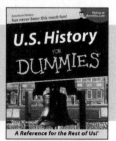